T0123035

THE EVOLUTION OF THE HUMAN PLACENTA

The EVOLUTION *of the* HUMAN PLACENTA

MICHAEL L. POWER and JAY SCHULKIN

The Johns Hopkins University Press
Baltimore

Printed in the United States of America on acid-free paper
9 8 7 6 5 4 3 2 1

The Johns Hopkins University Press
2715 North Charles Street
Baltimore, Maryland 21218-4363
www.press.jhu.edu

Library of Congress Cataloging-in-Publication Data

Power, Michael L.
The evolution of the human placenta / Michael L. Power and Jay Schulkin.
p. cm.
Includes bibliographical references and index.
ISBN 978-1-4214-0643-5 (hdbk. : alk. paper) — ISBN 1-4214-0643-8 (hdbk. : alk. paper)
1. Placenta. 2. Placenta—Diseases. I. Schulkin, Jay. II. Title.
RG591.P69 2012
618.8—dc23 2012006339

A catalog record for this book is available from the British Library.

Special discounts are available for bulk purchases of this book. For more information, please contact Special Sales at 410-516-6936 or specialsales@press.jhu.edu.

The Johns Hopkins University Press uses environmentally friendly book materials, including recycled text paper that is composed of at least 30 percent post-consumer waste, whenever possible.

To the many obstetrician-gynecologists who have collaborated with us over the years

They have informed us and improved our scientific endeavors with their knowledge and dedication to the science of women's health

CONTENTS

Introduction 1

1 The History of Placental Investigations 24

2 The Evolution of Live Birth in Mammals 50

3 Comparative Mammalian Placentation 73

4 The Evolution of the Human Placenta 96

5 Sex and the Placenta 119

6 Genes, Genetic Regulation, and the Placenta 138

7 The Placenta as a Regulatory Organ 163

8 Modern Gestational Challenges 196

Conclusion 223

References 231

Index 255

THE EVOLUTION OF THE HUMAN PLACENTA

Introduction

Both of us authors are parents. We have experienced the joys (as well as the other emotions) of creating new life and nurturing and guiding that child to become an independent adult. Of course, our wives had a great deal to do with all of this as well. Indeed, at the very beginning they did almost all of the work! We are placental mammals. The reproductive biology of placental mammals is strongly asymmetric with respect to investment in offspring. Mothers gestate them and nurse them, transferring large amounts of maternal resources to their offspring. The importance of fathers varies among mammalian species. In some species they provide sperm and little else. In other species—and we like to believe that human beings are an example—fathers provide considerable resources important to the success of their offspring; these resources do not come from the fathers' own bodies, unlike the case for mothers, who grow their offspring from nutrients drawn from their own tissues. Human biology is fundamentally affected by this asymmetry. The evolution of our species cannot be understood without reference to the different roles and constraints of male and female reproductive biology. In this one aspect at least men and women are intrinsically different.

This book is about a fundamental organ involved in human reproduction: the placenta. The placenta is the active interface of the most biologically intimate connection between two living organisms. The human placenta connects a mother and her baby physically, metabolically, and immunologically. Made from fetal cells, it is the first organ any mammal ever makes. That's right, the placenta is not a maternal organ; rather, it is formed from what are called the extraembryonic fetal membranes. All mammals, male and female, made a placenta at the beginning of their life. Besides being the first, the placenta is also a highly unusual organ—the only one ever to be connected to another individual (with the exception of conjoined twins). In essence, it links a mother and her fetus into a maternal-placental-fetal unit. The placenta regulates and coordinates the metabolism, growth, and development of both the mother and the fetus. The placenta grows and matures, taking on new functions as appropriate during gestation. Then, at the culmination of its function with the birth of the infant, it effectively orchestrates its own death.

The biology of the placenta is an immense topic, one that could not be covered in a single book. We have selected from among the many fascinating aspects of placental biology to create a coherent and informative narrative. This book is about the evolutionary biology of the human placenta. As evolutionary biologists, we explore how and when features of the human placenta arose, attempting to distinguish phylogenetically older aspects, that might represent constraints on human placental form and function, from more recently evolved characteristics, that might represent adaptive changes relevant to what makes us human. However, a placenta has been a fundamental adaptation for the extant (living) eutherian mammals (also known as placental mammals) for more than 100 million years. The existence of a placenta, and the intimate connection between mother and offspring that it represents, has influenced multiple aspects of mammalian biology, especially in the realm of biological regulation, from metabolism to the genome. It has required adaptive changes in immunological signaling for the mother to tolerate the allogenic (i.e., not-self) placenta and fetus. It has enabled regulatory signaling between the mother and her offspring during fetal growth and development. Mothers and babies are "speaking" to each other at a fundamental biological level from the moment of implantation. The intrauterine environment has been shown to have a strong influence over growth and development and to have significant effects on the future health and well-being of offspring long after birth.

All parents influence their offspring. In some species the influence is largely restricted to genetic contributions from both maternal and paternal sources. Sperm combines with eggs, eggs are deposited into a safe (we hope) part of the environment, and then offspring must develop to a stage at which they can obtain their own resources. Such is the reproductive strategy of most amphibians, fish, and many reptiles. The amniotic egg, which arose more than three hundred million years ago (see chapter 2), allowed greater maternal resources to be stored in the egg, thus allowing longer embryonic development time but also presenting more opportunities for mothers to influence the growth and development of their offspring by means of signaling molecules. For example, the steroid hormone content of bird eggs influences chick development (Sockman and Schwabl, 2000). But there is only a short time in which the resources and signaling molecules can be put into a bird egg; after laying, parental input is limited to body heat, an admittedly important and possibly even significantly variable input. The placental connection that evolved in the extant eutherian mammal lineage has increased the duration, amount, and flexibility of the maternal-fetal exchange. In addition, the placenta is a two-way connection, enabling fetuses to influence their mothers.

The existence of this intimate connection has increased the salience and selective pressures inherent in the inevitable to and fro of maternal-fetal and maternal-paternal genetic cooperation and conflict regarding fitness imperatives. Male and female mammals are fundamentally different in their reproductive biology, and this difference has profound evolutionary implications. The placenta is a primary arena where maternal and paternal genes interact, in both cooperation and conflict. The placenta appears to have allowed or even driven changes in aspects of placental mammal genetic regulation so that they differ fundamentally from those of other vertebrates such as birds or amphibians. For example, imprinted genes appear to be a mammalian adaptation within vertebrates (see chapter 5). Imprinted genes are important to placental development and function, and the placenta as a reproductive adaptation was a selective pressure for the evolution of gene imprinting (Reik and Lewis, 2005). The placenta appears to have influenced our biology at all levels. Accordingly, we also explore the placenta as an evolutionary force, one producing unique selective pressures that have affected mammalian biology, including our own.

Placentas Are Not All Alike

The complexity and diversity of placental function combined with the necessary communication between mother and fetus suggests that it is highly unlikely that any single placental architecture could have evolved as the best for all mammals; mammals are simply too diverse. There is considerable variation in placental form and structure among mammals, even within higher taxonomic units. For example, within primates, the anthropoid (monkeys, apes, and humans) placenta is of a completely different maternal-fetal barrier type from that of the prosimians (lemurs and lorises, but excepting tarsiers). Even within anthropoids, the platyrrhine (New World monkeys) placenta has significant structural differences from the catarrhine (Old World primates) placenta (Benirschke et al., 2006). And it is only among the great apes (including humans) that implantation is interstitial, with the fertilized egg (at the blastocyst stage) burrowing into the uterine epithelium.

The adaptive, functional differences among placental types can be difficult to judge. But that is exactly the task we have set for ourselves. We explore the characteristics of the human placenta and ask in what ways the uniqueness of the human placenta relates to the uniqueness of human beings. We examine how the uniqueness of other species might relate to their unique placental attributes. The placenta may be the organ that has undergone the most extensive and strong selective pressure. The result has been a myriad of solutions to the same basic problem: how to nourish and grow a fetus inside the mother until it can be born.

An Evolutionary Perspective

This book is written from the perspective of evolutionary biology, but what exactly does that mean? Some evolutionary biologists study anatomy and fossils, focusing on how the structure and function of animals' bodies have evolved. Some study the molecular changes of life over time, focusing on the variation and divergence of peptides and DNA over thousands of generations. There are theoretical evolutionary biologists, who try to develop models of how selection works at all levels of organization: DNA, genes, and peptides and other gene products, as well as the individual. We appreciate the work of these scientists and will refer to their work constantly in this book.

We study the evolution of regulatory systems. Both of us have especially broad and eclectic interests in biology, and we have been involved in a diverse range of biological research. One of us (JS) has studied, among other things, the neurobiology of feeding behavior, fear, attachment, and the connection between mind and body. The other (MLP) has studied peripheral physiology and metabolism, from energy metabolism to digestive physiology to milk composition. What we share is an intense appreciation for the regulatory nature of life. Living things are not static. A living creature is a complex assembly of organs and systems that must function in concert. The regulatory mechanisms that coordinate the metabolic and physiological processes that enable life are a major focus of our research and scientific thought.

Regulatory biology encompasses the entirety of mechanisms that serve to coordinate organs and systems to create a viable organism; viable is interpreted in an evolutionary sense, which implies an ability to transmit genetic material to successive generations. In many cases, these regulatory mechanisms involve the synthesis, secretion, and eventual binding of bioactive molecules in a target tissue or at particular molecular sites. When we speak of regulatory biology, we are speaking about the transmission and regulation of information. We'll often refer to regulatory molecules (e.g., cytokines, peptides, steroids) as information molecules. These molecules act locally and at a distance to coordinate cellular, organ, and system-wide responses to challenges by exchanging information among these systems.

It is a given that successful human reproduction requires an appropriately functioning placenta. The placenta performs critical functions in providing nutrients and gases to the fetus and in transporting metabolic wastes from the fetus to maternal circulation to be excreted. In addition, the placenta performs an essential immunological function, protecting the fetus (and itself) from attack by the maternal immune system. The placenta has evolved, however, to perform many adaptive functions beyond these key, primal ones. We now know that the placenta is intimately involved in molecular regulation of the growth, development, and metabolism of the fetus and the mother. The placenta is, fundamentally, a regulatory organ. It is a remarkable producer of information molecules: hormones, cytokines, immune function molecules, and so forth (Petraglia et al., 2005). These molecules act locally as well as at a distance to affect the mother, the fetus, and the placenta itself, in essence exchanging information among these entities in order to coordinate the necessary physiological and metabolic processes required to produce a viable neonate. This regulation of information extends naturally into the genome. The placenta has been a selective force on information-regulating systems at all levels: physiological, immunological, and genomical. We view the placenta as a canonical example of the regulatory nature of biology, and, accordingly, both of us have been drawn into studying its function and evolution.

Why is it important to study the evolutionary history of the placenta? Many aspects of biology can be studied at what is termed the proximate level. In other words, how does some aspect of biology work; what are the mechanisms; what is essential and what aspects have redundant pathways; what does this piece of biology do? Function can be deciphered without resorting to an evolutionary explanation. An evolutionary perspective provides the context of how, when, and why a trait has arisen. Evolutionary analysis aims to explore what is called ultimate causation—not just what a biological system does, but how and why it arose. There are usually many solutions to any biological challenge. Different species may adopt the same, similar, or quite different solutions to similar problems. The evolutionary perspective is invaluable to understand and evaluate why some solutions are similar and others are different among species. Closely related species will often have similar solutions to biological challenges, in large part because they are, in effect, coming from the same place. The biology underlying the trait in question likely came from their common ancestor. The descendant species may be constrained in the range of solutions they can evolve, due to inheritance of the genes from that distant common ancestor. The further back in time two species last shared a common ancestor, the less likely that this form of phylogenetic constraint will operate. Species that diverged long ago may have evolved different solutions to a challenge because they evolved different genetic potentials after their divergence. Similar solutions found in divergent species can imply convergent evolution; in other words, the two species arrived at the same solution even though they came from different genetic directions, so to speak. And finally, closely related species may still have diverged in some aspects of their biology, due to the unique selective pressures and circumstances of the two species' evolutionary history, even though they started from the same place. All these varying outcomes give different perspectives on and insights into the importance and functional significance of a trait. Divergent traits among closely related species imply different selective forces in their pasts. Convergence implies that the trait has significant selective advantages over a broad range of biological adaptations or possibly that the different species have faced similar challenges, despite their phylogenetic distance from each other.

Humans

Human beings certainly are especially successful as a species. Human population growth has been and continues to be quite impressive. We will exceed seven billion people in early 2012. Our reproductive physiology would appear to be quite functional. However, a number of authors have commented that human gestation is fraught with many diseases and poor outcomes for both the mother and fetus that appear to be, if not unique to human beings, at least far more common in human gestation than in the gestation of other mammals. These conditions include high rates of early pregnancy loss, nausea and vomiting of pregnancy, or morning sickness, preeclampsia and related maternal hypertension syndromes, and preterm birth (Jauniaux et al., 2006). These difficulties appear to be rare or absent in other mammals but common, and in the case of morning sickness ubiquitous, in humans.

If it is true that some or all of these pregnancy complications truly are unique to human beings, then they likely derive from biological adaptations unique to our lineage. That is not to say that these conditions were selected for or have any adaptive value. They may represent side effects, so to speak, of other biological adaptations that make us human. A human neonate has certain unique features that suggest the possibility of placental adaptations to support the development of these features. Of course that is true of all neonates of all mammalian species, but in this book our focus will continually return to human beings and the human placenta.

Many of the pathologies and poor outcomes of human pregnancy are, at their core, pathologies or failures of the placenta. Early pregnancy loss, preeclampsia, intrauterine growth restriction, preterm birth, and many other gestational diseases all have placental components. Even morning sickness probably derives from placental hormones released into maternal circulation, though in this case this is not pathology but rather a necessary adaptive placental function that has unfortunate side effects on maternal metabolism. Some researchers have even assigned adaptive value to morning sickness, an interesting hypothesis that we will explore in the last chapter.

Organization of the Book

The biology of the placenta does not easily fit into a linear story plan. Biology is inherently multifactorial and multidimensional. This is both a challenge and a source of enjoyment in writing a biology book; so many paths could be explored. In this book we take a wide-ranging, some might even say spiraling, path in our exploration of the biology of the human placenta. Because we are evolutionary biologists, this book will be predominantly about the evolutionary biology of the placenta. We seek to understand the human placenta and its roles in human gestation, fetal development, and birth through the lens of evolutionary biology and by comparison with other species. We seek to explore both the evolution of the human placenta as well as the role of the placenta in mammalian (and more specifically in human) evolution—in other words, what selective forces have been created by the adaptation of the placenta? Accordingly, we cover topics in this book that at first glance (or even at second or third glance) appear to be far removed from the placenta. For example, in chapter 5 we review what is known about the various mechanisms for sex determination in vertebrate species and the evolutionary history of sex chromosomes. We consider the consequences of differing numbers of sex chromosomes in individuals; that is, what are the repercussions of mammalian females having two X chromosomes, while males have only one? This leads us to the phenomenon of X chromosome inactivation, where only one of the two X chromosomes in every cell of a female mammal fully expresses its genes, with the other X chromosome being largely (but not completely) silenced. In marsupials, it is always the paternal X chromosome that is inactivated; in placental mammals it is random. Not surprisingly, the phenomenon of X inactivation extends to the placenta. Interestingly, in some species (e.g., rodents), although fetal X chromosome inactivation is random, in the placenta the paternal X chromosome is inactivated, similar to the situation in all marsupial tissues. Does paternal X inactivation represent an ancestral state, from before the divergence of marsupial and placental mammals, that has been retained in placental gene regulation for some species? Are there placenta-specific adaptive advantages to paternal X inactivation? And what insight does X inactivation provide with respect to the exciting new developments in genetic regulation, such as gene imprinting?

In order to retain the ability to explore as many paths as possible without hopelessly cluttering the main story, we have placed some material in "boxes," self-contained sidebars that either give further details regarding some aspect of the biology under discussion or merely explore an interesting facet of biology or history that we think some, if not all, of the readers of this book will enjoy. For we sincerely hope that people will take pleasure in this book as they learn about the placenta. We both find great delight in science and in learning about the world; we hope to share that sense of joy through this book.

Themes of the Book

To make this book manageable both to write as well as to read, we have focused on five interrelated themes: the placenta as a regulatory organ; the cooption of the immune system to become a regulatory system in early gestation; the evolution of genetic regulatory mechanisms due to selective forces enabled by the existence of a placenta; the biological anthropology of the placenta; and biomedical aspects of the placenta that relate to the previous four themes. The first three themes listed all relate to regulation, a fundamental aspect of life, and are applicable to all placental mammals. In the last two topics we focus on human beings and the evolutionary biology of the human placenta.

Our main research interest in the placenta is in its role as a central regulator of maternal and fetal physiology. Accordingly, a primary theme of the book is the placenta as a producer of information molecules that work locally and at a distance to coordinate functions of the multiple organ systems that comprise the maternal-placental-fetal unit. Reviewing the extensive research literature of the biology of the placenta and of implantation has reinforced that in addition to the usual categories of information molecules involved in physiological regulation (e.g., peptide and steroid hormones) the immune system is also involved in regulatory physiology, especially during implantation and early gestation. Although the immune system originated very early in evolution to serve an especially specific adaptive function, defense against attack by nonself organisms, at base immune system molecules are information molecules. Ultimately, complex multicellular life is about coordinating and regulating information among organ systems in order to keep the internal milieu in a viable state in the face of external challenges.

The placenta is a unique regulatory organ in that it coordinates metabolism between two genetically distinct organisms. The placenta must balance changing and sometimes conflicting maternal and fetal physiology. For example, early in human gestation the placenta acts to protect the fetus (and itself) from the high-oxygen and -glucose content of maternal blood. Early gestational fetal nutrition occurs via uterine secretions (called uterine milk by the ancient Greeks), not directly from maternal blood. After the first trimester, when organogenesis is complete and fetal growth begins to accelerate, the direct connection to maternal blood is established. Then maternal metabolism changes, regulated at least in part by placental endocrine signaling, to become hyperglycemic and hyperinsulinemic, enabling the large flow of maternal resources to the fetus.

Regulation also extends to the level of the genome. Identical DNA sequences can result in different phenotypes because of regulatory mechanisms that modify how DNA is expressed. This is most easily seen in the different cell types within any organism. Cells from the liver, heart, and kidney are all genetically identical in terms of DNA sequence but obviously different in phenotype due to their specific patterns of gene expression. The placenta is the most variable organ across mammalian species. Think about mammalian hearts or livers; of course they differ among species, but not to the extent that the placenta varies. The only other organ to approach such a level of variation might be the brain. How much of that variation is due to differences in genes and how much to differential regulation of genes? And, on a deeper evolutionary plane, the placenta appears to have been a selective force, favoring mechanisms of genetic regulation in mammals that are rare or even absent in most other vertebrates.

Biological anthropology is the scientific study of the biology that makes us uniquely human and the evolutionary path by which it arose. We focus on two aspects of human biology in which placental function likely has significant effects and for which therefore we expect that the placenta was under selective pressure to adapt. The first is the large human brain; the second is the role of disease, parasites, and other pathogens in our evolution. We will examine the evidence to see whether any unique features of the human placenta can be identified that perhaps arose as adaptations for addressing these challenges.

A human neonate has a brain approximately the size of an adult chimpanzee's. The human brain triples in size by adolescence, so obviously much of the brain growth occurs during lactation and after weaning. However, the human placenta must support the growth and maintenance of a fetal brain that at the end of gestation is about twice the size of a neonatal brain of our closest relatives. And very early postnatal brain growth is likely dependent on resources stored in the neonate that of course came from the mother, through the placenta. Thus it is not surprising that researchers have closely examined human placental biology for adaptations to meet these potentially enhanced resource requirements. Are there unique aspects of the human placenta whose adaptive function is to support brain growth? Alternatively, was our lineage in essence pre-adapted to be able to grow a large brain due to the type of placenta we share with our closest relatives? We will cast a skeptical eye on the evidence that the human placental type (hemomonochorial and deeply invasive) is required in order to grow a large fetal brain.

Diseases, parasites, and pathogens have influenced our evolution in many ways. Extensive amounts of DNA from incorporated retroviruses exist in our genome, and some of these ancient retroviral genes have even been co-opted to serve our biology. But there have also been adaptations in response to the morbidity and mortality caused by pathogens our ancestors faced. For example, malaria has been an important selective force for many human populations. There is evidence that pathogen loads and morbidity and mortality due to pathogens were exceptionally high at times in our evolutionary past. The invention of agriculture and the domestication of wild species certainly created the conditions for not only high pathogen loads due to higher population density and the consequences of poor sanitation, but also for exposure to novel pathogens brought into human contact from the formerly wild animals now living among us. We will consider evidence that pathogen load may have influenced human placental function.

This leads naturally to the final theme, the biomedicine of the placenta. For both of us, biomedical research is an important aspect of our research and intellectual efforts. We also strongly believe that evolutionary and comparative perspectives contribute to understanding human disease. In the final chapter we discuss certain pregnancy complications and poor birth outcomes with an evolutionary perspective and in light of the previously explored placental biology.

Placental Investigations throughout History

In the first chapter we explore the history of human thought regarding the placenta, from ancient Egypt, to the ideas and wisdom of Aristotle, to the beginnings of modern science. We also consider the modern reemergence of placentophagy (consuming the placenta and placental products) for health and beauty. Search the internet for "placenta" and you will discover a wide range of placental products for sale, derived from a variety of animal placentas. Appropriate scientific testing has been applied to few if any of these products, but they have their ardent supporters.

We will primarily explore the medical and scientific ideas that have evolved, in a different sense of the word, about the human placenta. Not surprisingly, conceptual advances concerning the biology of the placenta have often followed technological advances. The earliest placental investigations had to rely on what could be seen by the naked eye. Later investigations improved understanding as visual technology improved, first with light microscopy, and then with the power of scanning electron microscopy. These advances in imaging technology allowed the investigation of placental anatomy at the cellular level.

Because we are a visual species, most of the early placental investigations relied on imaging technology. Progress in knowledge followed progress in the ability to see things. However, we have the technological capacity to devise means of extending our other senses, in the case of placental investigations our ability to chemosense. The use of radioisotopes allowed scientists to track the flow of molecules across the placenta. Advances in immunological methodology gave rise to assay techniques to measure the various information molecules produced by the placenta. These in turn led to what we and others have termed the peptide revolution, where seemingly a new, potent signaling peptide is discovered every week. In addition, they led to the understanding that all organs have some endocrine function. The heart, liver, skin, adipose tissue, and placenta all produce information molecules that can act locally and at a distance to signal other organs and body systems.

These new technologies have allowed scientists to move beyond anatomy and structure and begin to investigate the regulatory functions of the placenta. These investigations have extended to the level of DNA, bringing new insight into the complexity of regulation that enables life. The conception of the adaptive function of the placenta has grown beyond its nutritive and respiratory functions to encompass endocrine and immune system regulation; the placenta is far from being a passive transporter of maternal resources and fetal waste.

The Past Matters

We inherited our biology from our ancestors, just as all species do. Some of our biology is extremely ancient. For example, the placenta is formed from three membranes: the chorion, the allantois, and the amnion (for the moment we will ignore a fourth, the yolk sac). These extraembryonic membranes are the membranes of the amniotic egg, which arose over three hundred million years ago. The eggs of all extant amniotes (reptiles, birds, and mammals) have these membranes, but obviously some specific aspects of their functions have evolved differently in various lineages. Chickens and humans share these membranes and their importance for embryonic and fetal development. In chickens the membranes are encased in a highly mineralized shell; in humans the membranes are within the uterus, with the fused chorion and allantois in contact with maternal tissue.

In keeping with the evolutionary theme, the second chapter of the book considers the challenges of producing live-born young and the wide range of adaptations that have arisen over the roughly half billion years of evolution since the first complex multicellular living organisms came to exist on earth. Producing live-born young is not unique to mammals but is rather an ancient adaptation that has been successful in many taxa. The coelacanth, a fish often termed a living fossil because it appears anatomically similar to fossil specimens from the Cretaceous, produces live-born young. The earliest evidence for live birth extends back more than three hundred million years ago in fossils of both fish and marine reptiles. Fossils of 380-million-year-old placoderms, ancient armor-plated fish, have been found with small armor plates from the same species within the body cavity (Long et al., 2008). Because these internal armor plates show no bite marks or signs of digestion, they would appear to represent tiny embryos and not a cannibalistic snack of hatchlings.

True live birth requires internal fertilization of the eggs, which are then retained until a certain level of development is completed, after which birth occurs. The evidence from placoderms implies that sexual intercourse arose nearly four hundred million years ago (Long, 2011) and has been disappearing and re-evolving in different lineages ever since. Internal fertilization allows the egg to be protected from predators and inhospitable external conditions. Perhaps most importantly, it allows the embryos to develop and grow within a controlled environment. Live birth is not uncommon; many species of fishes, lizards, and snakes produce live-born young. However, live birth is not known for any bird species, even though they practice internal fertilization, nor is it known for any of the extinct dinosaur species, some of which are sister taxa to the bird lineage. Crocodilians and turtles and tortoises are the other taxa that are not known to produce live-born young. All of these taxa have evolved a hard shell around the egg. This provides the developing fetus with another form of protection from external forces but may preclude, or at least greatly reduce the chances for, evolving live birth.

To continue our exploration of the evolution of the lineages leading to placental mammals, we go back in time hundreds of millions of years to explore the divergence of tetrapod vertebrate lineages, especially the divergence between the sauropods (whose descendants are reptiles and birds) and the synapsids (whose descendants are mammals). The various mass extinctions of species on earth have shaped the current mix of taxa in today's world. We briefly explore the Permian-Triassic extinction of 250 million years ago, which ended the dominance of the synapsids and began the sauropod era, culminating in the age of dinosaurs. The end of the dinosaur age (except for birds) at the closing of the Cretaceous initiated the rapid radiation and expansion of mammals. Somewhere between those two events, the eutherian mammals diverged from the metatherian (marsupial) mammals and eventually produced the placental mammals. A chorio-allantoic placenta (the broadest term that describes the placentas of all placental mammals, from aardvarks to zebras) arose in this lineage sometime between 100 and perhaps 150 million years ago.

Comparative Placentation

To understand our current placental biology we must understand our past. Comprehending the growth, development, and function of placental tissues will give important insight into the etiology of gestational diseases. But we do not just want to understand the "how" of placental development and function. Certainly understanding the genetic and developmental mechanisms that operate to construct a human placenta is important, but a comprehension of why and from where these developmental patterns arose is also critical. There is value to studying the past, not just the relatively recent past when we began as a distinct species, but also the ancient past, when many of our basic biological adaptations and constraints first arose. Comparisons with other species are also useful. There are often many solutions to common biological challenges; investigating the similarities and differences between ourselves and other mammals can give additional insight beyond what we can learn by studying our own biology.

Modern molecular techniques have enabled a better understanding of the relationships among mammalian species and the timing of their divergences. In chapter 3, we examine the diversity of placental structure and form within the context of the newly understood phylogenetic relationships among mammals in order to explore possible models of the ancestral placental form in eutherian (placental) mammals.

A chorio-allantoic placenta (box I.1) unites us with other placental mammals, such as bears, deer, dolphins, and porcupines, and distinguishes us from the nonplacental mammals, such as the kangaroo, a marsupial, and the platypus, a monotreme. Although marsupials and montremes are like us in some ways (for example, they have hair and feed their offspring milk), they do not have a chorio-allantoic placenta that nurtures their fetuses. But even within the placental mammals, there is considerable variation in placental structure. All of the above-mentioned placental mammal species can be distinguished from each other by some facet of placental structure. Dolphins have a diffuse, epitheliochorial placenta, deer have a cotyledonary, epitheliochorial (or synepitheliochorial) placenta, and bears have a zonary, endotheliochorial placenta (see box I.1 and table I.1 for explanations and descriptions). Porcupines are rodents and thus have a discoid, hemochorial placenta. Anthropoid primates (monkeys, apes, and humans) also have a discoid, hemochorial placenta. In this aspect of placental morphology rodents are similar to us, but the rodent placenta differs from the human in other aspects of placental morphology.

These similarities and differences reflect, in part, phylogeny. The evolutionary divergence between primates and rodents occurred after the divergence from those lineages that led to carnivores and cetartiodactyls (deer are artio-

BOX I.1 **PLACENTAL TYPES**

The eutherian mammal placenta is derived from fetal tissues; it is genetically identical to the fetus, though of course its gene expression differs markedly. The placenta is formed from three extraembryonic membranes: the chorion, the allantois, and the amnion. These are the membranes of the amniotic egg. Amniotes include reptiles and birds, in addition to mammals. The placental-fetal compartment is thus a modified amniotic egg. The yolk sac is also part of the placenta; its importance is highly variable among species. The eutherian mammal placenta is a chorio-allantoic placenta, which means the chorion and allantois fuse. The chorion remains the outer layer of placental cells, the tissue that is in contact with maternal tissue. The allantois forms the umbilical cord. The amnion becomes the fluid-filled amniotic sac that surrounds the embryo/fetus. Marsupials have a choriovitteline placenta, which means it is primarily composed of the chorion and the yolk sac.

Eutherian placentas can be categorized in a number of different ways: by their physical form; by the number of cell layers separating fetal from maternal blood; by the type of interdigitation of maternal and fetal blood vessels; and by whether there is a maternal component, the decidua, that becomes permanently attached to the fetal tissues that make up the placenta. There are four main categories of physical form: diffuse, cotyledonary, zonary, and discoid (table I.1). Placental types are spread across the mammalian phylogeny, and sometimes the distinctions are not absolute. For example, the nine-banded armadillo has a placenta that is zonary in shape but includes structures that could be classified as cotyledons. Until the genetic mechanisms underlying placental development are better understood, the phylogenetic significance of the physical differences in placental form will be hard to evaluate, and thus the ancestral placental type difficult to deduce.

The same is true for the number of cell layers separating maternal and fetal blood. There are three major categories, with various subtle complexities added, as would be expected of an evolved system with such fundamental importance. In an epitheliochorial placenta, there is no erosion of maternal epithelium, so there are a total of six cell layers (three maternal and three fetal) separating the maternal and fetal blood. In an endotheliochorial placenta, the uterine endometrial epithelial cells are destroyed in the implantation process, and chorionic trophoblast cells of the placenta are in direct contact with maternal endothelial cells. In a hemochorial placenta, all three maternal layers are eroded, and the chorionic trophoblast is directly bathed in maternal blood, because the maternal endothelium is penetrated by chorionic villi. Thus in a monohemochorial placenta such as is found in human beings, fetal blood is separated from maternal blood by only the three fetal layers of cells.

Table I.1 The four types of placental forms in eutherian mammals

Placental Form	Description	Examples of Species
Diffuse	The placenta completely surrounds the fetus, and almost the entire surface of the fused chorionallantois attaches to the uterine wall.	Horses, pigs, hippoptami, dolphins, whales; among primates, lemurs and lorises
Cotyledonary	The placenta attaches to the uterine wall in multiple discrete locations; the maternal attachment sites are called caruncles, and the placenta sections that attach to the uterine wall are termed cotyledons. The two together form what are termed placentomes.	Ruminants such as sheep and cows have 75–125 placentomes; deer have 4–6 larger placentomes; sloths and armadillos have cotyledonary-like placentas; horses have what are called microcotyledons in discrete sections of their diffuse placenta
Zonary	The placenta forms a band around the fetus; in some species it completely encircles the fetus, while in others the band is incomplete or forms two half-bands.	All carnivores (e.g., canines, felines, bears, seals), and also elephants, rock hyraxes, and aardvarks. Dogs and cats have complete bands; ferrets and raccoons have two half bands
Discoid	The placenta has a round, flat, disk-like appearance.	Rodents, bats, tenrecs, giant anteaters, red panda, and anthropoid primates, including humans

dactyls and dolphins are cetaceans, but molecular, morphological, and fossil evidence support their being combined into a new mammalian order, the cetartiodactyla [Gingerich, 2005; Springer et al., 2005]). Thus the similarity between rodent and anthropoid primate placentas probably reflects common descent. Similarly, most carnivore placenta are zonary and endotheliochorial, again reflecting common descent. Among cetartiodactyls, the ruminants (cows, sheep, deer, and so forth) have diverged from the common pattern among other cetartiodacyls, but otherwise pigs, hippos, dolphins, and whales have similar placental structure: diffuse and epitheliochorial.

The placenta is probably the most variable of all mammalian organs (Mossman, 1937; Kaufmann, 1983). This isn't surprising, considering the importance of the placenta to successful reproduction. Given the wide variation among mammals in morphology, ecology, reproductive strategy, and overall life-history strategy, it would have been amazing if any single placental type were able to meet all the varying challenges inherent in producing such different offspring. Common descent cannot explain all of the variation in placenta structure among mammals. Many of the differences in mammalian placental structure represent adaptive change—variations that conferred fitness advantages and thus were preferentially present in successive generations.

In chapter 4 we focus on the primate lineage, more specifically on the lineage that led to human beings. As an order, primates are somewhat interesting in having two divergent basic placental forms. The prosimian primates (the galagos of Africa, lorises of Asia, and lemurs of Madagascar) have epitheliochorial placentas. Tarsiers and all the anthropoid primates (monkeys, apes, and humans) have hemochorial placentas. The cladistic evidence supports the hypothesis that the epitheliochorial placentas of lemurs and lorises represent a derived condition, convergent evolution to a placental form that is also found in cetartiodactyls (e.g., pigs and dolphins) and perissodactyls (e.g., horses).

In this chapter we describe human placentation, from implantation to birth. We note the aspects of this process that have been suggested to be unique to humans (or at least unusual) and which have been suggested as adaptations to traits that are specifically human. Many researchers have sought to explain aspects of human placentation as adaptations for growing our large brain. We cast a skeptical eye on this hypothesis and suggest alternative adaptive scenarios that could be explored.

Finally, the placenta can interact with the maternal immune system in another manner besides communicating and regulating the maternal immune response. The mother can pass, through the placenta, her disease history to her fetus. In a number of species (e.g., rodents and anthropoid primates), maternal immunoglobulins, specifically IgG, are transported across the placenta to the fetus. Information about the mother's disease, parasite, and pathogen exposure is transferred to her offspring. To what extent has the adaptive advantage of transfer of maternal immunological history influenced the evolution of the human placenta? Human placental evolution may have been greatly influenced by infectious agents, pathogens, and parasites. Co-opted retroviral genes certainly have had a major role. But other pathogens have also likely influenced human placental biology. One of the adaptive purposes of the syncytium, the multinucleated tissue that forms the outer layer of the human placenta, is to serve as a barrier. It is a barrier to high-weight molecules but also to pathogens. For example, human placental syncytiotrophoblast is highly resistant to infection by *Listeria monocytogenes* (Robbins et al., 2010). Perhaps interstitial implantation, in which the newly developing embryo is isolated from the uterine lumen by maternal epithelium, has a protection-from-pathogens function.

Sex

The placenta cannot be divorced from sexual reproduction, and sexual reproduction cannot be divorced from sex determination. In chapter 5 we examine the two basic mechanisms by which the sex of an organism is determined: environmental (e.g., temperature-dependent) and genetic (e.g., sex chromosomes). Mammals rely on genetic sex determination using sex chromosomes X and Y; female mammals are the homogametic sex (XX) and males are heterogametic (XY), though there are a few rodent species in which the Y chromosome has disappeared and males have a single unpaired X chromosome. In many nonmammalian taxa the heterogametic sex is female (e.g., birds, snakes, and many lizards). In these species males are designated ZZ and females are ZW. In general, the heterogametic sex chromosome (Y or W) degenerates over time and becomes considerably smaller than the homogametic sex chromosome (X or Z). This leads to a mechanistic difficulty during meiosis, since the two sex chromosomes cannot appropriately pair. Plainly, this difficulty has been solved, but one aspect of the solution is that the sex chromosomes become at least temporarily inactivated in the heterogametic gamete (sperm for XY and egg for ZW). For mammalian females (XX) the inactivation of much (but not all) of one of the X chromosomes is either continued or more likely reestablished during development. This is hypothesized to be an adaptation to solve the dosage problem of having two sets of genes expressing gene products as opposed to only one set, as in male mammals (XY). Since a single X chromosome must be able to express sufficient gene products for male survival, and excess can often be as deadly as deficiency, the theory is that having two expressed X chromosomes in a cell would probably be lethal. Interestingly, in marsupials the inactivated X chromosome is always the paternal chromosome. In placental mammals X inactivation is generally random among tissues; a female mammal is a mosaic of X chromosome gene expression, with some tissues expressing the maternal genes and others the paternal. However, at least in rodents, the paternal X chromosome appears to be preferentially inactivated in the placental membranes of female fetuses.

Our exploration of X chromosome inactivation leads us to consideration of the various molecular mechanisms that regulate gene expression, including silencing the expression of regions of DNA. The developmental program of the offspring (and its placenta) is inherited from the genomes of its parents. In mammals, imprinted genes, genes whose expression is restricted to only one of the parental copies, appear to be an important adaptive mechanism for fetal development. The placenta seems to be an organ in which imprinted genes play an especially vital part. We will discuss how genes are imprinted, and more importantly, why imprinted genes may be so much more critical in the placental mammals than they are in most other vertebrates.

The theory of parent-offspring conflict has been a profitable area of exploration in evolutionary behavior and physiology. The placenta is the earliest and perhaps most profound site of interaction among parental (paternal and maternal) and offspring genomes. A major function of the placenta is to regulate the transfer of materials (nutrients and gases, but also bioactive molecules, many of which have developmental functions) from the maternal to the placental-fetal compartment. The fitness priorities of the mother and her offspring are not always completely aligned. The placenta is in a prime position to be engaged in maternal-offspring conflict. But even more intriguing, the placenta is an organ in which maternal and paternal genes can interact, both cooperatively and in competition. It is these interactions among the maternal-paternal-offspring genomes that provide the selective pressures and adaptive function for gene imprinting.

Genes and Genetic Regulation

In chapter 6 we continue to explore genetic regulation. The placenta arose from genetic potential that is quite ancient. As mentioned above, the extraembryonic membranes of the placenta derive from the membranes of the amniote egg (yolk sac, amnion, chorion, and allantois). The placenta provides a unique window into genetic and evolutionary mechanisms. The phenomena of gene imprinting, DNA methylation, gene duplication, differential regulation of genes by multiple promoters, co-option of existing genes for novel functions—all the mechanisms of genetic regulation and evolution—are on display in the placenta. And sometimes those genes aren't even originally from a mammal! A significant proportion of the mammalian genome, for example, as much as 8% of the human genome, consists of incorporated retroviruses (Lander et al., 2001). In many cases these genetic sequences from ancient retroviral infections have been silenced, either by DNA methylation or more permanently by the incorporation of nonsense mutations (Sugimoto and Schust, 2009). But in a number of instances, retroviral genes are expressed (e.g., as the proteins syncytin 1 and 2; Blaise et al., 2003) and have been incorporated into placental structure and function—an evolutionary irony in which the host has co-opted the genetic machinery of a past pathogen.

The Placenta Is an Endocrine Organ

Placental anatomy and structure are not the primary focus of this book; rather, placental regulatory function and evolution are. Anatomy is certainly important to function, and similarities and differences in placental anatomy certainly inform our understanding of placental evolution. But anatomy is only one piece of biology. We view the placenta as an endocrine organ and will focus more on its regulatory nature. We see parallels between the brain and the placenta in many placental functions. Indeed, the late, eminent reproductive endocrinologist Samuel Yen referred to the placenta as the "third brain" in pregnancy (Yen, 1994).

In the independent organism (i.e., after birth) the brain acts as the central controller of physiology, metabolism, and obviously, behavior. Before birth, the placenta has a substantial capacity to respond to the diverse conditions of the uterine environment. The fetal brain is immature and may not be able to perform all the normal, necessary regulatory functions that it will eventually perform post partum. The placenta synthesizes and releases a wider array of information molecules (hormones, cytokines, immune-function molecules) than probably any other organ, except possibly the brain. The placenta coordinates maternal physiology with fetal development. In many ways, the placenta acts as a go-between for the maternal and fetal central nervous systems. It serves as a central regulator of maternal-placental-fetal metabolism. Placental signaling influences the function of all maternal and fetal organs, perhaps most vitally the brain. Importantly, the maternal brain and the placental-fetal unit both require a stable source of oxygen and glucose. This implies that these fundamental organs of evolutionary success will often have competing, and even conflicting, demands that must be coordinated and accommodated through mutual signaling mechanisms. The placenta-brain axes may be the most critical understudied aspects of pregnancy for understanding both adaptive and pathological phenotypes.

In chapter 7 we explore this aspect of placental function and evolution. We also discuss in more detail the evolution and function of potent information-molecule families that perform important regulatory functions in the human placenta, such as the prolactin and growth-hormone families. These hormones derive from ancestral molecules that existed half a billion or more years ago. These molecules exist in all vertebrate species, but their functions are not identical in different lineages. Existing biology has been co-opted over time to produce novel functions and adaptations that distinguish the different types of vertebrates alive today. This concept is a recurring theme throughout the book; existing capacities are expanded and/or modified to produce novel biological adaptation.

Humans have retained unique aspects of anthropoid primate biology. Besides the basic anatomical similarities among anthropoid primate placentas, there are a number of signaling systems in the placenta that we will describe that appear to be unique to anthropoid primates. For example, in primates the growth hormone gene has been duplicated to produce a complex of genes that are predominantly expressed in the placenta. These produce the placental lactogens. Interestingly, rodent placentas also produce placental lactogens, but the rodent genes are derived from duplications of the prolactin gene. Human chorionic gonadotropin (hCG) is another anthropoid primate–specific hormone produced by the placenta. The known functions of hCG are to maintain the corpus luteum and induce progesterone secretion as well as possibly assist in maternal immune tolerance toward the implanted embryo. It also may be involved in morning sickness. The evolution of hCG has fascinating implications for understanding the selective forces affecting our lineage. Finally, anthropoid primates are the only species known to produce corticotropin-releasing hormone (CRH) in the placenta. CRH is a key element of a maternal-placental-fetal feedback loop that induces steroid production from the fetal adrenal, resulting in placental production of the estrogens necessary for successful gestation. In humans (and apes), maternal circulating CRH increases exponentially after the first trimester and then plummets to zero following the expulsion of the placenta after birth. Early and above-normal rises in circulating CRH are associated with preterm birth in humans, leading some to hypothesize that CRH is linked in important ways to the timing of parturition—a gestational clock, in effect. Our research has shown that the human and ape pattern of maternal circulating CRH across gestation is likely a derived characteristic, as it differs from the pattern found in New World and Old World monkeys (Power and Schulkin, 2006).

Placental Immunology

The immune system is ancient. The ability to distinguish self from nonself dates back to more than half a billion years ago. This is a fact obviously relevant to the challenges of live birth. Offspring are not genetically identical to the mother. In mammals that is inherently true, as the phenomenon of gene imprinting precludes parthenogenesis (see chapter 5). A mammalian embryo is nonself, from the maternal immune system perspective. If an embryo is retained within the mother, then it must be protected in some way from the maternal immune response. In the case of placental mammals, the placenta is the active interface where the maternal immune system and the fetal antigens would interact. The placenta is genetically identical to the fetus and thus nonself to the maternal system. In order for the developing fetus to be retained within the mother, there must be some kind of mechanism to inhibit maternal immunological responses and deter the mother's immunological system from rejecting the foreign entity her fetus represents. Clearly this problem has been solved by numerous species, our own included. However, many of the diseases of human birth have an immunological axis. For example, both preeclampsia and recurrent spontaneous abortion are associated with an increased inflammatory response and have been linked to an immune response to the foreign expressed genes from the father's genetic contribution to the fetus.

The immune system is, at base, an information system. Immune function molecules are information molecules. During human gestation, information that helps protect the neonate after birth is passed from mother to child. But in the earliest phases of pregnancy, during implantation, appropriate immune signaling has to occur between conceptus and mother. The placenta does not hide from the maternal immune system but rather signals it, and the maternal immune system becomes the placenta's partner in modifying uterine tissue to support successful implantation. The maternal and fetal immune systems have evolved a regulatory system that coordinates implantation, the initial development of the placental bed, and the maternal perfusion of placental villi.

In-Utero Origins of Disease

Finally, an area of considerable current research interest is the fetal origins of disease hypothesis. In this hypothesis, many of the chronic diseases of adults in the modern world have, at their base, risk factors that occur during gestation. For example, cardiovascular disease is greatly increased in men who were born small for gestational age but were subsequently raised in an environment of nutritional plenty. There are many axes by which these gestational risk factors can be transduced; in many the placenta will have considerable activity, especially via signaling from mother to fetus. The placenta must interact and communicate with most, if not all, maternal organs. Adipose tissue is just one example, but given the effects of maternal obesity on fertility, pregnancy, and birth outcome, and the dramatic increase in maternal obesity over the last few decades, the interactions between the placenta and adipose tissue are a timely and instructive area to explore.

Modern Gestational Challenges

In chapter 8 we examine a set of gestational conditions, from morning sickness to preeclampsia, in light of the materials presented in the previous chapters. The chapter starts with the early gestational conditions of early spontaneous pregnancy loss and morning sickness, which are negatively associated. The importance of fetal signaling in the maintenance of pregnancy will be examined in light of maternal-fetal conflict theory and the evolutionary advantages and disadvantages of early pregnancy loss. How might early pregnancy loss serve maternal fitness objectives? Does early pregnancy loss represent an adaptive maternal screening device? What might be the signals that influence whether an embryo is nurtured or shed?

Morning sickness is considered protective (as a risk factor) against pregnancy loss. The causes of nausea and vomiting of pregnancy are not well understood; they have been linked with human chorionic gonadotropin (hCG), which would associate them with maintenance of early pregnancy, though not in a causal way. Proposed evolutionary explanations for morning sickness will be critiqued.

Overarching all of these discussions is a consideration of the high incidence of early pregnancy loss in humans. We argue that what has been termed a reproductive inefficiency in humans compared to other mammals (such as rabbits and rodents) actually represents an evolved adaptation that has resulted in greater reproductive fitness.

1

The History of Placental Investigations

Humans have long speculated about the significance of the maternal-placental-fetal connection (Crofton Long, 1963; Jones and Kay, 2003). The placenta is viewed in different ways by different peoples. In modern Western culture it is primarily viewed medically. It is the afterbirth, sent to the pathology department possibly to be examined, especially if there was some complication with the birth or a poor outcome for the neonate. After some amount of time it will be disposed of, possibly incinerated, but it might be used in research or even sold to a pharmaceutical company. In other cultures the disposal of the placenta has ritual significance. It may be buried or burned in accordance with some form of funeral rite. In some cultures the placenta is considered a sibling or alter ego of sorts to the infant (De Witt, 1959). Considering that the placenta is derived from fetal tissue and is genetically identical to the neonate, that idea has some biological validity. Certainly in many cultures the placenta is merely considered the afterbirth—a problem if it is not expelled from the mother's body, but otherwise something simply to be discarded.

This is a science book, so we mostly discuss the biology of the placenta. Science is a way of knowing the world. It is a powerful and successful method of understanding, but it is not the only way. For much of human existence science has not been the way people tried to comprehend the world. We briefly explore the magic of the placenta, mainly as a means toward elucidating the history of human interpretation of the placenta as it has unfolded over the last few thousand years of scientific exploration. Our exposition is necessarily extremely broad and mainly oriented toward construing how past conceptions of the placenta influenced the development of scientific thought and investigation.

A key concept in this chapter is that technological advances have enabled conceptual advances; our progress in understanding the placenta has depended on the scientific tools we use to study it. At first human beings had only their own senses—sight, touch, smell, and taste—with which to understand the placenta. And to be honest, our chemosensing abilities are not highly developed; most of our early understanding of placental biology came from visual inspections. Aristotle was able to make some quite acute observations using just his naked eye. With the advent of lenses, and thus microscopes, scientists were able to look at tissue structure, eventually down to the level of the cell. The scanning electron microscope enables scientists to look within cells and at structures of cellular membranes. The development of laboratory animal models, starting with the rat, enabled experimental scientific methodology to be deployed in investigating placental function. Finally, the genomics and proteomics revolution is now being extended to the placenta. The placenta has been recognized as a signaling organ, producing a multitude of bioactive molecules that affect maternal and fetal metabolism and physiology. Technology has progressed to allow researchers to measure these peptides, steroids, cytokines, and so forth, and to examine gene expression and regulation at the DNA and RNA level. These aspects of placental biology are the focus of our book.

The placenta is attached to the infant and to the mother. That is a trivial statement but one that is fundamental in terms of understanding the evolution of the placenta. The placenta has biological meaning only in the context of a connection between the mother and the unborn child. This connection has also produced other, nonscientific, conceptions of the function and meaning of the afterbirth.

Before we begin our exploration of the modern biology of the placenta, we examine a few of the different interpretations of the placenta that have been documented by cultural anthropologists and historians. We look back in time as far as the ancient Egyptians and Greeks and then travel forward to modern ideas of the placenta worldwide. The placenta has and is considered to represent many things: nutrition, magic, medicine, and even a cosmetic product to enhance beauty and retain youth. Famous women from history such as Cleopatra and Marie Antoinette apparently used beauty products made from placentas. A casual search of the internet will uncover a variety of placental products for sale, mostly derived from horse and deer placentas but also from pigs, sheep, and other mammals. These products are to be consumed (usually in pill form these days) or are creams and lotions to be spread on the skin or massaged into hair (table 1.1). Few if any of these products have undergone rigorous scientific testing. However, they have their ardent advocates.

Placental Magic

The placenta has had ritual significance for a long time. This is not surprising given human beings' proclivity to seek understanding and meaning in most of

Table 1.1 Placental products

Product	Product Description	Placenta Type	Uses	Marketing Technique
Placenta 360000	Health drink	Pig	Preserve youth and beauty	Health, beauty
Placenta 10000	Face mask	Pig	Improve skin condition Rejuvenate skin for younger / healthier-looking skin	Beauty
Hask Placenta Super Strength Hair Conditioner	Hair repair treatment	Animal	"Protein diet for bleached, tinted, damaged hair"	Beauty
Perb	Placenta capsule	Sheep	Anti-aging supplement Wrinkle reducer	Health, beauty
Placenta Serum	Face serum	Pig	Anti-aging serum for face and body	Beauty
Maylande Royalé Soap	Face and body soap	Human	Replenish moisture in skin Unclog pores Soften and clarify skin	Health, beauty
Konad Iloje Flobu Whitening Care Line	Skin whitening treatment	Pig	Restore damaged skin Brighten and soften skin Protect skin from wrinkling and aging Improve complexion	Health, beauty

Medical treatment	Massage liquid	Human/horse	Help speed recovery of damaged tissue	Sports injuries
Food	Lasagna, spaghetti, cocktail, pizza, meat stew	Human	Hormones in placenta help speed maternal recovery from birth Provide nutrition to the body Potential stem cell benefits	Health Food, recipes
Greens' Twin Teddy Kit	Sustainable teddy bear	Human	Sustainable way to preserve a baby's placenta without having to eat it	Environment, green marketing
La Placenta Full Necklace	Necklace	Human	Preservation of one's baby's placenta	Fashion Religion, spirituality
Placenta encapsulation	Capsules/pills made from one's own baby's placenta after birth	Human	Balance hormones Enhance milk supply Increase energy Prevent PPD Allow for quicker recovery after birth Increase postpartum iron levels Shorten postpartum bleeding	Health/nutritional supplement
Koru Naturals and Natural Life Body Lotions	Body Lotion	Sheep	Replenish moisture in extremely dry skin Improve skin softness	Beauty

the common and less-common occurrences in their lives. The distinguished Scottish classicist and social anthropologist Sir James George Frazer wrote a pioneering work on religion and magic across human cultures (*The Golden Bough*, originally published in 1890). He included examples of ritual and magical beliefs from many cultures concerning the placenta and umbilical cord. For instance, he wrote that in Berlin the midwife would deliver the dried umbilical cord to the father with a strict injunction to keep it safe, for as long as it was kept the child would thrive and be free of sickness.

Placental magic can be categorized as either contagious or imitative. Contagious magic is the belief that objects that have been connected retain some affinity, even after the connection is broken (Frazer, 1890). Imitative or sympathetic magic is the notion that things that are somehow alike can be used to influence each other. The most familiar example to a modern audience might be the voodoo belief that a doll made in the image of an individual and containing hair, finger nails, or other parts of the body from that person can be used to affect that individual. Damage inflicted on the doll will result, at the least, in pain for the person represented by the doll. This can be thought of as either contagious or sympathetic magic. From this perspective it is fairly obvious how the placenta could be thought to have significance to both the infant and the mother, since it was produced inside the mother and was connected to both beings.

The earliest physical evidence of a cultural belief in the magical or spiritual significance of the placenta comes from ancient Egypt. Mummified placentas, human and animal, have been recovered from tombs. Although it cannot be proven (except perhaps by genetic testing), the assumption is that a human placenta found in an adult's tomb is the placenta from the birth of that adult. This implies that the afterbirth was carefully kept throughout life to accompany its owner into the afterlife.

In ancient Egypt, standards were carried before the monarch during processions. In keeping with the divinity of the kings (later pharaohs) of Egypt, the standards were associated with various gods and goddesses. Most standards could be easily identified as stylized animals or other obvious natural objects that were associated with a particular deity. One standard was puzzling; it looked like a bilobed bag or bladder with a cord attached. The standard was associated with the god Khonshu, who is linked with fertility, among other things. An intriguing theory is that this standard represented the king's placenta (Seligman and Murray, 1911). The placenta was considered the king's stillborn twin, and as the king was blessed with divinity, his placenta was in effect raised to the level of divinity as well. Interestingly, this ancient Egyptian ritual resembles the custom of the Baganda of Uganda—the placenta of a prince of royal blood will be preserved and carried in processions (Crofton Long, 1963).

A strong connection between an individual and their afterbirth is a consistent feature of many cultures, including African ones that may have predated the kings of Egypt. As stated in Frazer (1890), the Shilluks, a pastoral people of the White Nile, worshiped the spirits of their deceased kings. Each king was buried in the village of his birth, which was also the place where his placenta was buried. Burial of the afterbirth occurs in many cultures, often with rituals similar to those used for the burial of an infant.The Santal people of India refer to their birth village as the place where their placenta is buried (Jones and Kay, 2003).

In Korea, the placenta of a child of the royal family was buried in a special placenta jar to ensure good health. The placenta container consisted of a small jar in which the placenta was placed and a larger jar inside which the small receptacle was sealed. One of the numbered National Treasures of Korea (number 177) is a Buncheong placenta jar currently housed at the Korean University Museum.

The Caul

There are three extraembryonic membranes that comprise the placenta: the chorion, the outer layer of the placenta that is in direct contact with maternal tissue; the allantois, which forms the umbilical cord; and the amnion, which becomes the amniotic sac, surrounding the fetus and containing the amniotic fluid, which protects the fetus. In rare instances a child is born with part of the amniotic sac still clinging to its body, often on its head, and occasionally on its face. This occurrence and the amniotic tissue often have been considered to have great significance. In English it is termed a caul. English-language literary references to the caul date to the mid 1500s. A Scottish variant of the term (kell) appeared in print around 1530 (Forbes, 1953). The word *caul* and its Scottish variant *kell* are sufficiently similar in spelling and pronunciation to *cowl,* as in "monk's cowl," that the words might be cognates. Obviously the physical similarity between a caul that drapes over an infant's head and a monk's cowl further suggests a relationship and gave rise to a belief that children born with a caul were destined for the monastery or convent (Forbes, 1953). Forbes (1953) also relates an instance in which a child was born with a caul that was put aside with little initial notice of its condition. Several hours later the caul was found to have the words "British and Foreign Bible Society" impressed upon it, a discovery that engendered great excitement over such a seeming miracle. The presiding doctor's account was more prosaic; he determined that the caul had been placed on a bible on which those words were embossed.

Medically, there is little cause for alarm if an infant is born with a caul, other than the need to quickly clear it from the nose and mouth so the baby can breathe. However, in many cultures, past and present, this occurrence has significance. Cauls were sold to witches and sorcerers. They were used as ingredients in love potions. People preserved them; often they wore them in charms or amulets (Forbes, 1953). There were many superstitions surrounding cauls, both concerning a child born with one and also about the potential power and uses of cauls themselves.

In England, cauls were sold to sailors as protection against drowning (Frazer, 1890). This myth of protection is actually common. In many cultures, children born with a caul are thought to be safe from death by drowning. In England there are records of cauls being offered for sale to mariners from the 1700s through the first world war. The first line of the fourth paragraph of Charles Dickens' novel *David Copperfield* is "I was born with a caul, which was advertised for sale, in the newspapers, at the low price of fifteen guineas." The passage goes on to speculate whether the failure of any acceptable offers to appear was due to a shortage of money or a shortage of faith among sea-going people. The changes in price for these charms most likely reflected demand, as supply was presumably relatively constant. Times of war, especially naval warfare, were associated with higher prices for cauls. During times of peace the price of cauls decreased significantly (Forbes, 1953).

Cauls have been thought to act as charms for many other aspects of life besides safety from drowning. They have been used as love charms. They have also served to assist a person in their public speaking; a caul was supposed to confer the gift of eloquence (Forbes, 1953). Cauls would be sold to lawyers, who believed possessing a caul would aid them in pleading their cases.

Historically, many famous or powerful people were said to have been born with a caul: Alexander the Great, Napoleon, and Lord Byron, among others. Undoubtedly there were many more ordinary people than extraordinary ones born with cauls, but myths and legends concern the famous (or infamous) people.

The Placenta in Ancient Greece

Where did science begin? That is a fascinating question that we will not be able to answer in this book. However, in ancient Greece a scientific attitude and approach was in evidence. The first known recorded words from an incipient scientific perspective regarding the placenta were by an Ionian physician, Diogenes of Apollonia, circa 480 BC. He was the first person known to believe that the placenta was the organ of fetal nutrition. The competing theory was that the fetus was nourished by sucking from "uterine paps" formed by the uterine lining. This latter view was continued by the Hippocratic School (De Witt, 1959) and led to this incorrect notion surviving well into the Middle Ages.

Aristotle (384 BC–322 BC) could be considered the father of natural philosophy, the study of the natural world. Aristotle studied many aspects of biology—what we would call botany, natural history, anatomy, embryology, and taxonomy. His writings provide the first surviving recorded systematic studies of zoology. He dissected many animals, but never humans. Accordingly, his writings concerning the anatomy of humans are far less accurate (and indeed largely wrong) compared to his writings on other animals. Regarding other species, he was remarkably accurate in many cases. For example, he described the hectocotyl arm of male cephalopods (e.g., squid and octopi), which males use to store spermatophores and then to fertilize the females' eggs. His description of this anatomical structure was largely disbelieved by future readers and considered to be fiction until the nineteenth century, when Cuvier rediscovered the hectocotyl arm.

Aristotle dismissed the Hippocratic theory of uterine paps, correctly noting that the fetus was enclosed in membranes and thus could not suckle from the uterine wall. He arrived at a theory of embryonic nutrition based on his comparative work. His studies of chick embryos led him to suggest, by analogy to the yolk sac, a nutritional function for the placenta.

Although in ancient Greece midwifery was considered to be the exclusive realm of women, Aristotle did write about some aspects of birth. When a person is born, her or his birth is shortly followed by the expulsion of the placenta from the uterus. If this doesn't happen there are serious maternal health consequences. If the afterbirth is retained it will lead to infection, bleeding, and eventually maternal death. This fact is well understood by all human cultures or at least by those individuals who participate in the birth process. The ancient Greeks wrote about the methods midwives used to extract a recalcitrant placenta. According to Aristotle, if the placenta were not naturally expelled from the uterus, the infant with the umbilical cord still attached would be placed on a water-filled bladder that slowly leaked. The woman would sit or squat above the bladder, and as the bladder emptied, the weight of the infant slowly sinking toward the ground would exert a continuous tension on the placenta through the umbilical cord, encouraging it to be expelled (figure 1.1).

Aristotle may be best known in zoology for devising a classification scheme for animals. Aristotle classified dolphins as mammals, not as the fish that most people of his time thought they were. His insight was based on multiple things that he observed about dolphins: they breathed air and had lungs, they produced milk to feed their young, and they produced a placenta, like other mammals. Aristotle correctly recognized this fundamental connection among the eutherian mammals.

Galen (129-217?) was born in the city of Pergamun in Asia Minor. He went on to become the physician to several Roman emperors and was a brilliant anatomist who made extensive dissections of animals, including monkeys. Because Roman law forbade human dissection, Galen based his ideas of human anatomy on his dissections of other species, especially Barbary macaques. His ideas about human anatomy, disease, and medicine were extremely influential for the next 1,500 years. He synthesized the works of Hippocrates and Aristotle but sided with Aristotle by stating that the embryo is nourished by direct connection with maternal blood through the umbilical cord (De Witt, 1959). The Galenic concept that the maternal and fetal blood circulations were somehow connected lasted for 1,500 years or more and caused much controversy and conflict among anatomists. It wasn't until the work of John and William Hunter in the 1700s that the theory of Galen was definitively proven incorrect and that maternal and fetal blood supplies were shown to be separate.

The Placenta during the Renaissance

The Renaissance in Europe combined technical, scientific, artistic, and philosophical rebirth and reawakening, but more importantly, progress in all these human endeavors. They were not as separate as they have come to be today.

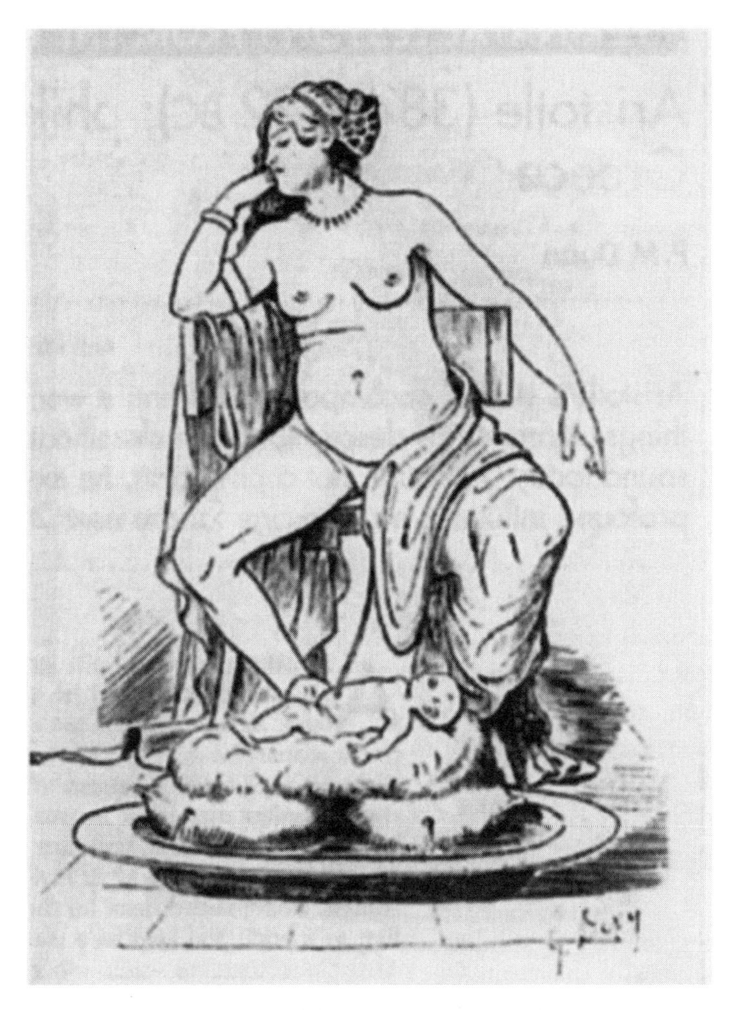

Figure 1.1. Depiction of an ancient Greek midwife technique for expelling the placenta after birth. Reproduced from *Arch Dis Child Fetal Neonatal Ed*, "Aristotle (384–322 BC): Philosopher and scientist of ancient Greece," PM Dunn, vol. 91, F75–F77, copyright 2006, with permission from BMJ Publishing Group Ltd.

Artists had to be technicians and even scientists. They were working with and developing new materials and methods. The greatest of them, such as Leonardo da Vinci and Michelangelo, were dissecting animal and human corpses to determine how muscles, tendons, and ligaments worked together to make a functioning body. Advances in anatomy were important in art, science, and medicine. In truth, this is not so different from today, but in those days it was frequently the same person making the discoveries and advances in all these fields of endeavor. Science, art, philosophy, and often religion were intertwined.

The modern science of anatomy had its beginnings in the 1500s. Galen had been considered the leading authority on anatomy since his death, and for over a thousand years, much of anatomy was taught as an affirmation of the Galenic conception. However, during the 1500s the writings of the classical Greeks—Aristotle, Galen, and others—were reexamined. Anatomical dissections were performed and the results compared with the words of the great thinkers of the past, especially Galen. Andreas Vesalius (1514–1564) and his students at Padua University in Italy led this new way of thinking that relied on observation and experiment. Vesalius pointed out the facts, long overlooked, that Aristotle and Galen had been prohibited from performing human dissections and that their ideas about human anatomy were largely based on analogy with animals, in Galen's case mainly the Barbary macaque. In 1543, at the age of 28, Vesalius published his book *De humani corporis fabrica libri septem,* which many mark as the beginning of modern anatomy (Adler and Adler, 1970). Based on his and his students' dissections of human beings, this book corrected many errors in the anatomical ideas of Galen. Vesalius was attacked by the Galenists, including his own teacher, Jacobus Sylvius (1478–1555) of France. He quit his position at Padua University in disgust, but his students and colleagues carried on the work of developing the modern science of anatomy; his student Gabriele Fallopius (1523–1562) discovered the Fallopian tubes, and his colleague Renaldus Columbus (1516–1559; box 1.1) appears to have been the first person to use the word *placenta* in print (Adler and Adler, 1970; Longo and Reynolds, 2010). Fallopius became the chairman of the Padua University anatomy department and remained in that position until his death, after which his student Fabricius ab Aquapendente (1537–1619) became the chair. Fabricius is considered by some to be the father of the science of embryology; he conducted an extensive study of embryo development and created a set of illustrations of the human gravid uterus and placental membranes (Longo and Reynolds, 2010).

A famous illustration of a gravid uterus with late-term fetus and attached placenta by Leonardo da Vinci (1452–1519) is remarkable for both its detail and for being fundamentally incorrect; Leonardo depicted the human placenta as cotlydenary, like that of a cow or sheep. Like Leonardo, Vesalius published an incorrect anatomical drawing of the human fetus connected to the placenta in the first edition of his book, depicting the placenta as zonary, as in dogs.

BOX 1.1 THE WORD *PLACENTA*

The human placenta has been and continues to be called many things: the afterbirth, second birth, secondyne, and uterine liver, among others. The word *placenta* was first applied in print to the human organ in the sixteenth century by the Italian anatomist Renaldus Columbus, also known as Matteo Renaldo Colombo, a colleague of Andreas Vesalius at Padua University in Italy. He used the word in his life's work, *De re anatomica*, published a few months after his death in 1559. Although this work is little known nowadays, it was a significant advance in the study of anatomy in its time (Ghirardini, 1982). Renaldus Columbus was the first European anatomist known to have described pulmonary circulation well, and in doing so he disproved the ideas of Galen.

The twelfth book (of fifteen) of *De re anatomica* concerned the formation of the fetus and the situation of the fetus in the womb. In it, the author refuted the Hippocratic ideas of fetal nourishment (via uterine paps) and another idea of the time, that the fetus consumes menstrual blood. Instead, he held that the fetus receives nourishment through the umbilical cord. He correctly concluded that the amniotic fluid serves to protect the fetus, though he incorrectly decided that it derives from fetal sweat. He agreed with Aristotle that the vertebrae are the first elements of the skeletal system to appear, but he disagreed with Aristotle's statement that the heart is the first organ, correctly identifying the nervous system as the first to develop.

The word *placenta* comes from the Latin word for flat cake, which in turn is derived from the Greek word 'plakous' (from 'plax,' meaning "flat"; plural 'plakountes'), which referred to a Greek flat bread or cake. Exactly what type of flat cake 'plakous' referred to is uncertain, as there likely were regional and community differences in this food. Some plakountes appeared to be flat bread sprinkled with herbs, garlic, and onion during baking, perhaps a distant ancestor of pizza. Plakous was also a baked concoction of cheese and flour, sweetened with honey (Wilkins and Hill, 2006). No recipe for Greek plakous survives; however, a recipe for Roman placenta is found in *De agri cultura,* or *On Agriculture*, written by Marcus Porcilius Cato (Cato the Elder) around 160 BC. This complex recipe combines multiple layers of dough and cheese, sweetened with honey—an ancient version of cheesecake. In all cases placentas and plakountes were round and flat, resembling in shape the biological human placenta. Despite the fact that most mammalian placentas do not resemble a flat cake of any kind, the term has stuck.

He corrected the drawing in the second edition to correctly depict the human placenta as a single disc. During this time the Galenic idea that there was a direct connection between maternal and fetal blood supplies was reexamined. Leonardo concluded that maternal and fetal blood supplies were separate, as did Vesalius. Giulio Cesare Aranzi (Arantius; 1530–1589) dissected pregnant ruminants, sows, and dogs and concluded that Galen had been wrong. Others of that day disagreed, however. Andre Du Laurens (Laurentius; 1558–1609) reasserted the ideas of Galen, and even Fallopius's student Fabricius ab Aquapendente attacked Arantius for doubting Galen (Longo and Reynolds, 2010). In the latter case, though, the disagreement was due to Fabricius having misinterpreted the cotyledon crypts in the sheep placenta, which match openings in the uterine caruncles, as being the openings of blood vessels from the placenta matching up to maternal blood vessels, thus indicating a connected blood supply. In this case disagreement was not based on deferring to the authority of Galen but rather to interpretation of observations. Even though Fabricius was incorrect, he was following the tenets of the new science of anatomy.

Fabricius attracted many talented anatomy students to Padua University, including William Harvey (1578–1657). Harvey completed the work of others (e.g., Columbus) from Padua University by correctly describing the pulmonary circulation. He used the experimental method to show that blood flowed in a circle through the body, and that the heart, not the liver, was the engine driving circulation. He also explored fetal circulation and determined that maternal and fetal circulations were separate. He then asked a simple but profound question (Harvey, 1847, 263): "How does it happen that the foetus continues in its mother's womb after the seventh month? Seeing that when expelled after this epoch, not only does it breathe, but without respiration cannot survive one small hour; whilst, as I have before stated, if it remains in utero, it lives in health and vigour more than two months without the aid of respiration at all." His conclusion was that the placenta served as the fetal lung as well as the fetal liver (Dunn, 1990) but, as oxygen had not yet been discovered, Harvey could not determine what required substance was supplied by the placenta or how it was supplied (Longo and Reynolds, 2010).

In the 1600s, the concept of the placenta as an organ that transferred necessary substances from mother to child was established. Some considered that food or nutrition was transferred to the fetus via the placenta. Others, such as Harvey, suggested that the placenta served as the fetal lung, and still others thought the placenta functioned as a liver, purifying the blood.

John Mayow (1634–1679) of Oxford showed that some part of air was required for both burning a candle and for animal respiration. He termed this fraction of the air "nitro-aërial particles" and, in answer to Harvey's question, held that fetal blood must somehow contain this substance. The placenta must provide "nitro-aërial particles" in addition to "nutritious juices." He argued that the placenta should be thought of as the uterine lung instead of as the uterine liver. In the 1700s, the work of Joseph Priestley (1733–1804), Antoine Lavoisier (1743–1794), and Carl Scheele (1742–1786) resulted in the discovery of oxygen. This discovery enabled Lavoisier to extend and confirm the concept suggested earlier by Mayow that combustion (fire) and respiration (energy metabolism) were similar. Oxygen, an invisible, odorless component of the air we breathe, was now recognized as necessary for life. The question posed by Harvey, how does a fetus live within its mother?, could now be refined to, how does a fetus obtain oxygen? The placenta became an obvious candidate.

Although it was clear that material necessary for life came from maternal blood and somehow passed to the fetus, there was still disagreement among experts about whether the maternal and fetal blood supplies connected or were separate. Wilhelm Noortwyck (ca. 1712–1778) was the first person known to have injected uterine vessels with material to trace the blood vessels; unfortunately, he incorrectly concluded that some of the vessels containing the injected material were fetal in origin (Longo and Reynolds, 2010). However, the technology to resolve the issue now existed. Colored material (e.g., wax or liquids) could be injected into uterine arteries and veins of dissected material so their paths could be traced. The brothers William (1718–1783) and John (1728–1793) Hunter used this technique to investigate placental circulation by injecting colored waxes into various arteries and veins. William demonstrated the convoluted nature of the uterine arteries (spiral arteries) that deliver blood to the placenta. Using the wax technique, they also showed that maternal and fetal circulations were kept separate. The brothers fell out over who should have precedence for this discovery and were only reconciled shortly before William's death (Longo and Reynolds, 2010). Despite the magnificent illustrations in William Hunter's published obstetric atlas, *The Gravid Uterus,* the question of separation of maternal and fetal circulation was still considered unanswered by some. However, injection experiments continued to uphold the Hunter brothers' discovery. Fillipo Civinni (1805–1844) confirmed the separation of maternal and fetal circulations by injecting colored liquids into the placentas of humans and animals from dissected cadavers.

During this period, knowledge became driven by innovation and new techniques. Replication of results, rather than reputation, had become the standard for judging ideas. Galen could still be admired, but his incorrect ideas were discarded.

Despite the controversy over maternal and fetal blood circulation, one aspect of placental biology had now become rather settled. The placenta was understood to have a role in transferring vital components from the mother to her fetus. Erasmus Darwin (box 1.2) was familiar with the works of John and William Hunter; indeed, he had attended William Hunter's anatomical lectures as a student. Erasmus Darwin combined Lavoisier's theory of oxygen and metabolism with the anatomy findings of the Hunter brothers in his major work, *Zoonomia,* in which he devoted a chapter to the placenta. He proposed that a major function of the placenta was the oxygenation of fetal blood. He likened the function of the placenta to that of the lungs or the gills (Dunn, 2003). Interestingly, and incorrectly, this insight led him to propose that the amniotic fluid, not the placenta, was the source of fetal nutrients (Pijnenborg and Vercruysse, 2007). The advance due to the discovery of oxygen inadvertently led to a temporary retreat in knowledge regarding the source of fetal nutrition; in essence, the placenta was underestimated. Providing oxygen to the fetus was considered such a vital function (and it certainly is) that the idea that this is merely one of many functions performed by the placenta was not immediately grasped. Single-theme hypotheses have often been favored in human thought, which perhaps tells us far more about the limits of the human mind than it does about how the world really works.

Interestingly, many of the scientists of that day had what could be called an incipient evolutionary conception of biology, which was quite evident in their discussions of the placenta. Erasmus Darwin certainly considered similarity in form and function to be indicative of common descent (Dunn, 2003). There was no theory of mechanism yet; that would come later, from his grandson, Charles Darwin. But there was a phylogenetic awareness in the attempts to understand placenta function and form.

Although Charles Darwin was well aware of his grandfather's work, there is no evidence that Charles Darwin considered the placenta in formulating his theory of evolution. "Darwin's bulldog" Thomas Huxley (1825–1895), however, proposed that the structure of the placenta of different mammalian species provided important information about their evolutionary descent. Huxley published a detailed account of the rat placenta in support of his opinion con-

BOX 1.2 ERASMUS DARWIN AND THE LUNAR SOCIETY

Erasmus Darwin was a well-respected physician and philosopher. He was also the grandfather of Charles Darwin and of Francis Galton, two of the leading scientists of their generation. He was a founding member of the Lunar Society, whose members were a veritable Who's Who of European and American science and technology. Members of the society included Benjamin Franklin, Joseph Priestley, and James Watt. Another founding member was Darwin's friend Matthew Boulton, a self-taught scientist and metal worker by trade. Erasmus Darwin's theoretical knowledge and Boulton's practical expertise complemented their shared love of experimentation. Boulton eventually partnered with James Watt to produce steam engines for industry, and he brought Watt into the group. The Lunar Society was extremely informal; indeed, because it had no bylaws or membership list, some argue that it wasn't truly a society. Yet it was extremely influential in the development of science and technology from its start in 1765 until 1813. The name of the society derived from the decision to hold meetings each month around the time of the full moon. This choice was made in order to facilitate traveling home at night, given the lunar illumination.

cerning the close phylogenetic position of rodents relative to humans (Pijnenborg and Vercruysse, 2004). Huxley focused on whether at parturition there was loss of maternal tissue when the afterbirth was expelled, in other words, whether or not the placenta was invasive and became inextricably intertwined with uterine tissue (Longo and Reynolds, 2010).

The Placenta and the Development of Modern Science

The anatomical understanding of the placenta advanced as microscopic and serial-sectioning technology improved. Again, advances in knowledge were driven by advances in techniques and technology. The structure of the placenta could now be examined at the level of cell layers. In the late 1800s, Matthias Duval (1844–1907) demonstrated the invasion of human maternal uterine tissue by placental tissue (Pijnenborg and Vercruysse, 2006). The Dutch embryologist Ambrosius Hubrecht (1853–1915) was highly influenced by Darwin's theories of evolution and Huxley's ideas regarding insectivores as the basal mammalian type. He studied implantation in the hedgehog in detail. He proposed the term "trophoblast" for the cells forming the outer layer of the blastocyst, in light of the seemingly obvious nutritive significance of these cells that were in direct contact with maternal tissue. Interestingly, the fact that embryologists had now taken a leading role in placental investigations resulted in the embryonic suffix being applied to even the mature cells of the placenta (e.g., *syncytiotrophoblast* instead of *syncytiotrophoderm*). An embryological view of the placenta perhaps obscures the fact that although the placenta may be embryonic in origin, it develops and matures into a fully functional, if transient, organ.

The technique of embedding tissues in paraffin for later sectioning was introduced by Edwin Klebs (1834–1913) in 1869. A further important technical advance for studying anatomy came in 1886, when Charles Minot (1852–1914) invented the automatic rotary microtome, capable of cutting precise, ultra-thin sections of tissue. Using the microtome, Minot produced a detailed account of the microscopic structure of the human placenta. With such improvements in microscope technology, the late nineteenth and early twentieth century saw rapid advances in describing placental tissue. Hans Strahl (1857–1920) completed many comparative studies, starting first with carnivores, for which he described hemophagous regions in dog and ferret placentas (Carter and Mess, 2010). Strahl also published several papers on primate placentas, showing that the greater bushbaby (Strahl, 1899) and two lemur species have epitheliochorial placentas (Strahl, 1905), quite different from human beings and other anthropoid primates. Emil Selenka (1842–1902) collected gravid uteri of gibbons and orangutans; after his death the material went partly to Strahl. The papers by Selenka and Strahl generated from these organs were the first documentation of interstitial implantation and the formation of a decidua capsularis in great apes (but not gibbons), similar to the human condition (Carter and Mess, 2010).

Otto Grosser (1873–1951) made substantial contributios to the study of placental anatomy. He received his initial training in medicine but preferred the anatomical study. In the early 1900s, he became chairman of the department of anatomy at the German University in Prague, where he was instrumental in developing that department into one of the most famous of the era. He published two monographs on the histology of the placenta (Grosser 1909, 1927). In these texts, he proposed the classification of placental structure in terms of the number and type of tissue layers separating maternal and fetal blood. The Grosser classification (in modern terminology: epitheliochorial, syndesmochorial, endotheliochorial, and hemochorial) is still used today, though it has been extended and modified by subsequent researchers (e.g., Enders,

1965). For example, the term *syndesmochorial* is no longer considered useful, and placentas of this type are now classified as epitheliochorial or sometimes synepitheliochorial to recognize that placentas of this type have cells that fuse with maternal epithelial cells to form hybrid multinucleate cells (Wooding and Burton, 2008). In broad terms, the Grosser classification is useful for dividing placentas into three types:

1. epitheliochorial placentas, which have all six layers of tissue between maternal and fetal blood intact (three fetal layers, consisting of endothelium lining the allantoic capillaries, connective tissue, and chorionic epithelium; and three maternal layers, composed of endometrial epithelial cells, connective tissue, and endothelium lining the endometrial blood vessels)
2. endotheliochorial placentas, in which the maternal epithelial cells and connective tissue have been lost, but the maternal endothelium remains
3. hemochorial placentas, in which all three maternal tissue layers are lost

Some of the findings Grosser used for his classification scheme were compromised by the limits of light microscopy. The technology was not sufficient to visualize the tissue with the resolution needed to differentiate the placental types correctly. The advent of electron microscopy allowed superior visualization of the cell layers and resulted in modifications of Grosser's classification scheme. His basic categories are still recognizable, however, and are as appropriate a system to classify placentas from mammalian species as any other yet proposed. The functional and evolutionary significance he attached to the different types of placentas has been shown not to be well founded, however (Wooding and Burton, 2008).

Although the work of Grosser advanced the study of comparative placentation, his views regarding evolution and functional morphology that were not well supported served to confuse rather than illuminate the evolutionary history of the placenta. His evolutionary view was one of a progression from primitive to advanced, in which humans stood at the pinnacle of phylogenetic attainment (Kaufmann, 1992). He also believed that the decrease in tissue layers separating maternal and fetal blood resulted in an increase in efficiency of placental nutrient transport. In his view, humans had a more efficient placenta than did "lower" mammals. Although there is some truth to this notion that the loss of maternal tissue layers is a way to increase material transfer among the mother, the placenta, and the fetus, it is most certainly not the only way.

Based on the size and metabolic needs of fetuses produced by species with epitheliochorial placentas, we see that this type of placenta, too, can be enormously efficient. For example, horses have a diffuse, epitheliochorial placenta. That type of placenta is about as different from a human placenta as possible. Newborn horses are quite large and well developed, with sufficient musculature to be able to walk soon after birth. Gestation is not particularly long; thus the transfer rate of resources from mother to fetus is certainly not low compared with humans. Although horses certainly do not have particularly large brains, newborn horses have more and better developed muscles than do human babies. Like the brain, muscle is metabolically expensive. So the relative metabolic demand on the mother is not obviously different between humans and horses, after accounting for their different body sizes.

For nutrients that require a transport mechanism to cross cell membranes (e.g., sodium ions) the number of tissue layers is indeed a key determinant of transfer rate. For gases (e.g., oxygen and carbon dioxide), the thickness of the tissue barrier is what matters, not how many layers of cells exist. The thickness of the tissue layer can certainly be decreased by reducing the number of cell layers, but epitheliochorial placentas have regions where the tissue barrier has thinned greatly, without loss of cell layers.

These two assumptions by Grosser, that human beings represent a pinnacle of evolutionary progress and that fewer cell layers imply greater efficiency, led to the conclusion that the epitheliochorial placenta type was likely to be the ancestral form and that a hemochorial placenta is a derived, evolutionarily advanced state. This conclusion does not withstand scrutiny based on modern biological knowledge, as discussed in chapter 3.

The Modern Anatomists

Harland Mossman (1898–1991) examined the anatomy of the placenta from a developmental and comparative perspective and produced an influential book, *Vertebrate Fetal Membranes* (Mossman, 1987). His work associated morphology with function. He demonstrated that maternal and fetal circulation in the labyrinth of rodent and rabbit placentas flowed in opposite directions (Mossman, 1926). Such a counter-current flow would allow highly efficient exchange of respiratory gases. Elizabeth Ramsey (1906–1993) and colleagues used cineradiographic techniques to show that counter-current flow does not occur in the Rhesus placenta (Ramsey et al., 1960).

The limits of light microscopy had led to some incorrect conclusions regarding rabbit, rodent, and shrew placentas. It appeared that the trophoblast had been eroded, creating a hemo-endothelial placenta, with maternal blood in contact with fetal endothelial cells. Using electron microscopy, Allen Enders was able to demonstrate that layers of trophoblast persist at all times during gestation in rabbits and a variety of rodent species (Enders, 1965). Working with Wimsatt and Mossman, Enders was able to show that in the shrew placenta, a thin layer of trophoblast always existed between the maternal and fetal endothelial tissues (Wimsatt et al., 1973). Thus, the hemo-endothelial placenta has not been shown to exist; all suggested examples have been shown to retain a trophoblast layer throughout gestation.

With electron microscopy, the tools to investigate placental morphology on the cellular level were up to the task. Researchers with a comparative perspective, such as Allen Enders and Kurt Benirschke, published many examinations of placentas from multiple species. Enders did his PhD work on the mink, and one of the species he studied early in his career was the nine-banded armadillo, which he showed had a villous, hemochorial placenta (Enders, 1960). Armadillos exhibit a fascinating variation in reproductive strategy in that they produce identical quadruplets from a single blastocyst. Enders investigated the structure of the armadillo blastocyst and the process of implantation, describing the manner in which this occurs (Enders, 2002). Enders also investigated implantation and early trophoblast development in several nonhuman primate species, such as the baboon (Enders and King, 1991), the Rhesus macaque (Enders, 1989), and the common marmoset (Enders and Lopata, 1999).

Benirschke approached the study of the placenta from the perspective of a pathologist. His book (written with Peter Kaufman and Rebecca Baergen) *Pathology of the Human Placenta* (2006) is encyclopedic in its coverage of placental development and pathology. He has had a long association with the San Diego Zoo, which fits with his comparative interests. He maintains a web site (http://placentation.ucsd.edu/index.html) that includes his descriptions of placentas from the multitude of species he has personally examined.

Placental Physiology

In the early 1900s, physiologists began to investigate the placenta as an organ. Specifically, the mechanisms by which nutrients were transferred from the mother through the placenta to the fetus were explored. Louis Flexner (1902–1991) and colleagues used radioactive isotopes to study molecular transport in the placenta (e.g., Flexner and Roberts, 1939). They were not able to demonstrate the pattern predicted by Grosser, that fewer cells layers resulted in more efficient transfer. They did show that the transfer of water and sodium to the fetus increased throughout most of gestation but then dramatically decreased during the last four days prior to parturition (Flexner and Roberts, 1939). Thus the placenta regulates nutrient exchange. Morphology is important in regulation, but other regulatory mechanisms are of perhaps greater importance. Placental efficiency of transport of key molecules is not a simple matter of the number of cell layers nor of their thickness. The placenta is an organ of regulatory physiology; its function is not static and fixed by morphology but rather is under regulatory control.

The Placenta as an Endocrine Organ

We are a visual species. Much of what we learn about the world comes from our ability to see things. Therefore, it is not surprising that early explorations of placental biology relied on technological enhancements of visual images, culminating in scanning electron microscope pictures. This instrument allowed increasingly detailed examination of placental structure, but it did not particularly advance knowledge of placental biochemistry. It is an interesting thought experiment to consider how the history of placental investigations would have progressed if our sensory abilities were more like that of dogs, sharks, or other species with excellent chemosensory abilities. Upon examination of a fresh placenta, would such scientists have exclaimed over the richness and diversity of the chemicals contained in this organ? Would we have discovered the signaling and regulatory functions of the placenta earlier?

Thankfully, humans have the capacity to imagine and appreciate the importance of things that our own biology has difficulty detecting as well as the capacity to develop technologies that can detect them for us. We can't see oxygen, but in the 1700s people knew it existed and could measure it. And scientists were able to appreciate that there were secretions within the body that traveled through the bloodstream and affected other organs and overall organism behavior and function, even though they could not see these substances.

In 1656, Thomas Wharton (1614–1673) published *Adrenographia,* the first thorough account of the glands of the human body, including the first description of the thyroid. He also confirmed Harvey's account of the placenta and studied the umbilical cord, providing the first published description of the gelatinous connective tissue found within (Wharton's jelly), which gives the umbilical cord its resiliency and pliability (Üstün, 2003). Wharton's jelly has been shown to be a source of stem cells and possibly a reservoir of primitive stem cells produced early in development.

A little over one hundred years later, Theophile de Bordeu (1722–1776) proposed the idea of internal secretions; this was an elaboration of the Hippocratic theory of humors into the beginnings of the scientific conception of endocrine function. Claude Bernard (1813–1878) and Charles Edouard Brown-Sequard (1817–1894) furthered the concept, and, unlike Bordeu, added experimental results to demonstrate that substances produced by glands can affect physiology. Many of these molecules, such as insulin, epinephrine, thyroxine, vitamin D, estrogen, and testosterone, were isolated and characterized in the early-to-mid 1900s.

The tools of modern science now allow us to peer deeply into the biology of tissues and organs. Present-day biology has become more involved with regulation and metabolism, largely because technology has progressed to where we can measure the signaling molecules and their receptors. This ability has changed our conception of the endocrine system and of organ function. In 1983 a paper titled "The right auricle of the heart is an endocrine organ" was published; it documented the synthesis and secretion of hormones from heart tissue (Forssmann et al., 1983). Since then, every organ in the body has been shown to exhibit endocrine function—to be able to secrete information molecules as well as bind them. The body is a complex system of information transfer and regulation.

Perhaps most relevant to our modern understanding of placental function is the ability to probe the molecular signals involved in metabolism and endocrine function. We can now proceed past structure and begin to examine the molecular biochemistry that allows the placenta to perform its myriad functions. The placenta is seen to have additional functions besides nutrient delivery and waste removal. It is an endocrine organ, producing a wide array of information molecules that act locally on its own tissues and at a distance on both maternal and fetal tissue. The placenta produces as many or more of these peptides, cytokines, steroids, and immune-system molecules as any organ in the body except possibly the brain. We can now move beyond an examination of placental structure as it relates to function and extend our investigations to the role of the placenta in regulatory physiology. This aspect of placental biology is the primary focus of our book—the evolutionary history of the development of the placenta as a central processor of information for the maternal-placental-fetal unit.

Technology has advanced to a point where scientists can now examine both genomic structure and expression in tissues. Placental gene expression is obviously different from that of other organs; in some ways that statement merely reflects that the gene expression of all the different tissues within an organism has to vary—how else could there be different tissue types? But some specific aspects of the placenta's origin and function suggest that placental gene expression may have some unusual, if not unique, facets. Certainly there appear to be aspects of gene regulation (e.g., imprinted genes) that are common in mammals, but rare or possibly absent in other amniotes, that reflect the adaptive pressures presented by a placenta.

Suppression of Immune Rejection

Peter Medawar received the Nobel Prize for his work on graft rejection. He also published a provocative paper on viviparity (Medawar, 1953). Based on his immunological work, he raised the question, why doesn't the maternal immune system reject the fetus? Or more specifically for humans, since it is the placenta that is in direct contact with maternal tissue, how does the placenta interact with the maternal immune system to allow the flow of nutrients, gases, and wastes to and from the fetus?

The question is intriguing. The assumptions behind the question were not without errors, however. Medawar was considering the fetus as an allograft. In a graft, the cells are in direct contact with blood; arteries and veins must invade and grow into the grafted tissue, or it will die. The placenta is separated from maternal blood by a variable number of cell layers, depending upon the species. It is an analogous situation, especially for the invasive tissue of the human placenta, but many of the important factors that would cue immunological attack are either missing or greatly attenuated.

This is not to say that there is little or no maternal immunological response to pregnancy. Pregnancy is a state of moderate inflammation, and the immune system is intimately involved in implantation and remodeling of uterine tissue. Many of the pathologies of human pregnancy have aspects of immunological dysfunction. Immunological signaling is an important function of the placenta, but its context and function are different from what is usually thought of when referring to the immune system (see chapter 7). The maternal immune system acts as a partner with the blastocysts to support implantation. There is actually a lack of certain signals from the placental membranes, but positive signaling also regulates and informs the maternal response.

Placentophagy

Many (but not all) mammals eat the placenta after birth. Many theories exist about why mothers consume the afterbirth, ranging from removing evidence that might attract predators or unfriendly conspecifics to recycling the nutrients and other bioactive constituents of the placenta back into maternal metabolism. We also have a long history of human perception of placentas as an asset for health, youth, and beauty. There are modern advocates for a mother to consume the placenta after she gives birth.

It is somewhat interesting to consider the phrasing used by many proponents of placentophagy. The woman is urged to eat "her" placenta. Of course that phrasing is just as accurate as referring to the infant as "her" baby. However, the assumption appears to be that the woman is reingesting her own tissue, regaining the nutrients (and other bioactive molecules) she deposited during pregnancy. There certainly is some truth in that view—the resources that produced the placenta had to first pass through the mother. In one sense she did grow the placenta, in the same sense that she grew her baby. However, genetically the placenta is linked to her baby, not to her. The placenta is a fetal organ, not a maternal organ. She would be consuming fetal tissue. Perhaps it doesn't sound as appealing to say that a woman should eat the tissue, now separated from her baby, that formed her baby's first organ.

Modern Science and New Magic

The placenta continues to be viewed in magical ways by many people. It is associated with health and well-being, youth and beauty. For many it remains a potent symbol of the connection between mother and child. After most births the placenta is disposed of by the hospital, but a significant minority of mothers request to keep it. Some use it in ritual, for example, burying it and planting a tree on top. The child is then told that the tree is his/hers. There is even a kit that will turn a placenta into a teddy bear, perhaps the ultimate in a personalized gift from parent to child.

Other people eat the placenta, either cooked, or, in the most modern manifestation, cooked, freeze-dried, ground into powder, and placed in capsules, to be taken as a nutritional supplement. Recipes can be found on the internet for placenta lasagna or a placenta smoothie. Proponents of the placenta as a supplement suggest that it reduces the likelihood and extent of postpartum depression, enhances maternal energy after birth, and increases milk production. We are not aware of any clinical trials that have investigated whether ingesting placenta products after birth has any identifiable effects. However, there have been studies that indicate that eating the placenta and/or amniotic fluid affects the endogenous opiate system in rats (Kristal et al., 1985). Also, in the early 1900s studies examined whether women ingesting dried placenta affected milk composition and the growth of their infants (Hammett and McNeile, 1917; Hammett, 1918). Although the effects on milk production and nutrient content of milk were minor, infants of the women given capsules containing their dried placenta grew faster than control infants (Hammett, 1918). This led the authors to hypothesize that some factor produced by the placenta acted in an endocrine manner to enhance growth, and they suggested that during gestation the placenta, in addition to providing nutrients, may produce growth factors that influence fetal growth.

This early suggestion that the placenta had endocrine function was controversial. Editorials cautioned about extending the role of the placenta beyond nutrient, waste, and gas exchange into this new area of biology. Endocrinology was in its infancy at the time, and scientists lacked the tools to detect the wide range of endocrine function coming from most tissue. We now know that Hammett was correct that the placenta produces growth factors, although that does not verify his hypothesis that the observed difference in infant growth was due to a transfer of placental growth factors from the mother's digestive tract to her milk and then to the infant.

Placental products are not just for mothers, however. Afterbirth products are touted as health and beauty aids for others as well. There is an extensive market for animal placenta, too—horse, deer, pig, and other mammalian placentas are offered for sale or used in a variety of products for beauty and health. Drinks made from horse placenta are used as wellness and beauty aids. Shampoo made from horse placenta is marketed for lustrous, thick hair; as the consumer recommendation says on the web site, "there's nothing wrong with putting animal afterbirth in your hair." Lanolin cream with sheep placenta is marketed for dry skin. Bottled, blended pig placenta is sold in Japan for health and vitality. Deer placenta is advertised as the most potent of placental health products with the statement that deer are "higher-order" mammals, with no explanation of what that might actually mean.

Modern science has demonstrated that the placenta not only contains, but indeed produces, a litany of potent bioactive molecules. The contemporary, nonscientific conception of the placenta has incorporated this knowledge. Many of the placental products that can be purchased online, products that advertise their ability to promote health, youth, and beauty, now reference the existence of these molecules in the placenta to support the proposed function of their pills or lotions. The efficacy of these natural products is supposed to come from the potent hormones, nutrients, and other molecules produced by the placenta. Modern marketing uses the language of modern science.

Myths about the placenta exist today and have probably existed since our first ancestors were able to view the world and wonder about the significance of everyday events. They all contain an element of truth, given that truth is often an individual or cultural expression of understanding the world. In this book, we focus on scientific truths. Concerning the placenta, scientific truths have changed considerably over time. The placenta has not changed; rather, our knowledge and conception of what the placenta is, what it does, and how it evolved has changed. Much of that change in understanding is due to the development of more powerful tools of discovery, but scientific knowledge progresses. We continually test and improve our theories and concepts. This is the power of science, as well as the frustration, on occasion. Scientific truths rarely last forever, but when they fail, it usually advances our comprehension.

2

The Evolution of Live Birth in Mammals

Producing live-born young is not unique to mammals; it isn't even unique to vertebrates. Scorpions and other invertebrates, such as tunicates of the genus *Salpa* (salps) and some species of velvet worms, also produce live-born young. In these last two examples there is even a tissue connection between the maternal and embryonic forms in some species that transfers maternally derived nutrients to the offspring before birth—functionally a placenta, though not at all similar in cellular structure to the mammalian placentas that are the topic of this book.

A species can produce live-born young by two basic means: ovoviviparity or true viviparity. In ovoviviparity (also called aplacental viviparity) there is no embryonic connection with any maternal tissue. After hatching, the embryo must either rely completely on its yolk sac for nutrients (just as an embryo in an egg does) or obtain nutrients through feeding. In some species the embryo actually swallows substances it finds within its mother's reproductive tract and digests that material in its own digestive tract. These substances can be secretions from the mother or, in some cases, the embryo may consume its own siblings. The latter description of "ovoviviparity with fetal cannibalism" is a reproductive strategy that has been observed in at least a few species of live-bearing sharks.

With true viviparity, there is a direct connection between maternal and embryonic tissue, and maternal resources are transferred to the embryo via that connection. In viviparous species the maternal resources initially deposited into the egg are typically minor; that is, the eggs of truly viviparous species have very small yolks, whereas many oviparous species deposit a substantial amount of yolk into their eggs.

A Brief Evolutionary History of Live Birth

The oldest possible examples of live birth come from 380-million-year-old placoderm fossils. In simplest terms, several of these ancient armor-plated fish fossils have been found with a fossilized impression of soft tissue that has been interpreted as an umbilical cord connecting to a yolk sac (Long et al., 2008). Whether there was any more intimate connection to maternal tissue cannot be determined; these embryos were retained within the female reproductive tract, but there may not have been much, if any, transfer of maternal resources past that initially deposited into the yolk.

Among vertebrates, live birth is certainly less common than oviparity. However, live birth has arisen independently in many different vertebrate lineages. Fish, amphibians, reptiles (excluding archosaurs), and mammals all include species that produce live young. Of course, there are also species in all these groups, including mammals, that lay eggs. The mechanisms and adaptations by which species produce live-born young can vary substantially, and the evolutionary paths are many. It is estimated that there have been more than one hundred independent transitions from egg laying to live birth among living vertebrate taxa (Dulvy and Reynolds, 1997).

One proposed adaptive hypothesis holds that retaining the eggs and then the developing embryos within the mother's body is a protective strategy first evolved in aquatic species. For one thing, eggs deposited into the environment, and particularly the marine environments where ovoviviparity is thought to have first arisen, are at risk of being scattered over large areas due to currents, waves, and other water movements. For some species, allowing the eggs to be scattered is an adaptive strategy and spreads their offspring (and genes) as widely as possible, in hopes of achieving success. The safe habitats where embryos can develop successfully are limited, however, and of course eggs are often eaten by predators. The internal maternal environment can be one of the safest, though it does entail the proverbial all-eggs-in-one-basket downside. If the mother is eaten by a predator or dies for other reasons, she loses all her offspring as well. The fossil creatures from hundreds of millions of years ago that are known to have utilized live birth are all fairly large, probably apex, predators. The maternal environment thus would be fairly safe, at least from predation. However, this finding might reflect a bias—because fossilization may preserve internal embryos of large species more easily; perhaps there were smaller species giving birth to live young, but the features were less likely to fossilize.

The earliest evidence of live birth does not include any suggestion of placenta-like structures. This lack of evidence is not definitive, as such soft-tissue structures do not fossilize easily. Indeed, some live-bearing species have transient placenta-like structures that connect the egg to the mother, but these structures are lost after the embryo hatches within the mother.

External Fertilization and Live Birth

A key adaptation to allow live birth is internal fertilization. Many fish and amphibians utilize external fertilization of the eggs. The female lays the eggs and the male excretes the sperm to fertilize the eggs after they are expelled from the female's body. This practice would seem to preclude live birth. However, there are species that use external fertilization followed by gathering of the eggs into a specialized structure. Seahorses and pipefish are interesting examples of an external adaptation that allows a semblance of live birth. Males of these species have a brood pouch, an enclosed area on either the tail or ventrum. The female essentially lays her eggs in the male's brood pouch. The male releases sperm into the water and the eggs are fertilized in the pouch, where they grow and develop.

So, do male seahorses really get pregnant? That depends on the definition of "pregnant." Unlike the mammalian uterus, the seahorse brood pouch is not an internal organ. The embryos are separated from the male immune system by a barrier of skin. However, the male does regulate the brood pouch environment, providing oxygen and nutrients via vascularized tissue. Perhaps most important, the male regulates the osmotic environment in which the embryos develop (Partridge et al., 2007), reducing the salinity within the pouch to approximate blood level early in development and gradually increasing the salinity to that of seawater by the time the embryos have developed sufficiently to be "born." Embryos removed from the pouch ahead of time fare poorly in normal seawater, perhaps reflecting a difficulty in meeting the homeostatic challenge provided by the osmotic gradient (Azzarello, 1991). Functionally, the male seahorse uses a special feature of his anatomy to provide a safe, regulated environment in which his embryos can develop and grow.

What does the male seahorse gain from this parental effort that is far beyond the usual male vertebrate contribution? Complete confidence in paternity, for one thing. In many species with external fertilization, sneaky fertilizing males can swoop in and deposit their sperm into the egg mass. Not so for seahorses. There is no doubt that the embryos in his pouch are genetically his. Thus the costs he incurs from his parental care results in direct effects on his fitness. His embryos are protected from some predators, though obviously not from predators of adult seahorses. Another adaptive advantage is that the environment in which the embryos develop can be controlled. Environmental perturbations can have significant effects on embryonic development in all species, and early insults to the developmental process can have life-long consequences. This is the basis of the current hypotheses regarding the etiology of many chronic adult human diseases such as hypertension, cardiovascular disease, and type 2 diabetes. In-utero conditions are thought to predispose adults to be unusually vulnerable to these conditions (e.g., Barker, 1991). Finally, keeping the embryos in the brood pouch affords the male seahorse some ability to allocate resources among his offspring and potentially to recover some resources. Reduction in brood size does occur in some species, and in the broad-nosed pipefish a study of radioactively labeled embryos demonstrated paternal uptake of radioactivity (Sagebakken et al., 2010), implying that embryos that died were at least partly absorbed by paternal tissue.

These examples are fascinating reminders that nature produces a wide variety of successful strategies for living and reproducing. There are many solutions to life's problems. They also illustrate the fact that the outside world is a dangerous place for eggs and developing embryos. Adaptations to enable live birth, in all its varied forms, serve a protective function for embryos, providing a safer, more controlled environment in which to develop.

Amniotic Eggs

Humans are tetrapod vertebrates. So are frogs and salamanders, but not fish. The living tetrapod vertebrates (amphibians, reptiles, birds, and mammals; figure 2.1) are descendants of a lineage that diverged from the lobe-finned fishes as much as four hundred million years ago. A living nontetrapod descendant of those fish is the coelacanth, which has been called a living fossil. Interestingly, coelacanths produce live-born young. So, what distinguishes us from the amphibians (but not from reptiles and birds)? We are amniotes—that is, we produce amniotic eggs.

What distinguishes the amniotic egg from fish and amphibian eggs is a unique set of membranes: the amnion, the chorion, and the allantois. These membranes are multicellular, vascularized tissues that grow out of the embryo and thus are referred to as extraembryonic membranes. The amnion surrounds the embryo and creates a fluid-filled cavity in which the embryo develops. In humans and other placental mammals, this becomes the amniotic sac and is filled with amniotic fluid. The chorion forms a protective membrane around the egg. The allantois is a special membrane that performs gas exchange and stores metabolic wastes from the embryo; eventually it forms the

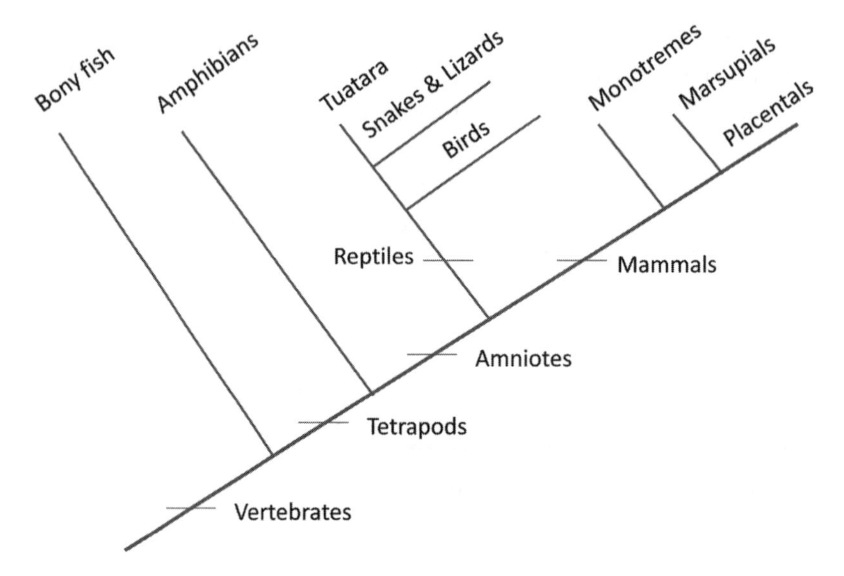

Figure 2.1. Basic phylogeny of vertebrates.

bladder. In placental mammals these two extraembryonic tissues, the chorion and the allantois, form the placenta and the umbilical cord, respectively. The antecedents of the modern mammalian placenta date back to adaptive changes in egg structure that occurred during the Carboniferous era, approximately 350 million years ago.

What were the adaptive advantages provided by amniotic eggs? The extraembryonic membranes of the amniotic egg protected the embryo from desiccation. The amniotic egg also allowed for greater transfer of maternal resources; eggs could become larger when they had the protective chorionic membrane. More maternal resources could be placed into the egg, and the embryo had greater protection from the external environment. The suggested water-loss prevention function of the amniotic egg matches with other adaptations in amniote biology that also appear to prevent desiccation. For example, the skin of amniotes is relatively impervious, contrasting with the skin of amphibians, through which water and gases easily flow. Amniotes possess a high density of renal tubules and a water-resorbing large intestine. Much of amniote biology can be viewed as adaptations to support fluid balance, which enables them to live on land, away from water.

The shell of eggs of early amniotes was most likely poorly mineralized (Laurin et al., 2000). This would explain why no fossil amniotic eggs have been found that date older than approximately 220 million years ago, though undisputed fossil amniotes are known from as far back as 310 million years ago. Eventually some taxa developed greater mineralization of the shell, which was more conducive to fossilization. A well-mineralized shell is a characteristic of the archosaurs (dinosaurs, birds, and crocodiles).

Although the amniotic egg is well adapted to dry land and indeed provided an adaptive advantage to the lineages of amniotes that colonized the land masses, the initial adaptive origins of the extraembryomic membranes are poorly understood. The fossil record provides evidence of the eventual success endowed upon amniotes due to the ability to lay their eggs on land, but it does not give us much insight into the origin questions. All modern amniotes still retain the extraembryonic membranes, but their uses vary among the different lineages. For birds, turtles, and many reptiles, the anti-desiccation advantage is still relevant. Birds have evolved a highly calcified shell, which is a further adaptive protection for their eggs. For mammals, the extraembryonic membranes serve different functions, as do those membranes of squamate lizards that produce live-born young nourished by a placenta. In these taxa, the extraembryonic membranes of the amniotic egg perform a maternal-fetal connection/communication function. Amniotic eggs are associated with internal fertilization, which allowed egg retention to evolve. Egg retention provides protection for the egg from inhospitable external conditions as well as from predators. So, did early stem-lineage female amniotes retain eggs within the reproductive tract for a longer time than nonamniotes? Was an original function of the chorion, amnion, and allantois related to increased egg development within the mother? Did early terrestrial amniotes all have eggs that resisted desiccation, or did some species solve this challenge by retaining their eggs within the reproductive tract to reduce the amount of exposure to the external environment? Certainly, these are not mutually exclusive hypotheses. The original amniotes may have retained eggs within the reproductive tract for a considerable time and then produced eggs that were resistant to desiccation. These are both adaptive advantages that would have benefited species shifting from a partially aquatic to a wholly terrestrial life.

Standard phylogenetic analyses are unlikely to be able to settle this issue, because the initial amniote adaptation occurred so long ago; too many adaptive changes have occurred since the original amniotes arose to make analysis feasible. The fact that there are no early amniotic eggs in the fossil record probably reflects their low fossilization potential and is not reliable evidence for an early adaptation to extended egg retention or to live birth. It is unlikely that early amniotes produced live-born young, but it is not impossible. Live birth predates the amniotes within the pretetrapod lineage—lobe-finned fishes produce live-born young. Live birth could be considered an adaptation to fully terrestrial living, providing further protection of embryos from an inhospitable environment.

Egg Retention

Development is affected by multiple factors. First, of course, is the genetic underpinning of the developmental program. Some of these genes must be quite ancient and essentially unchanged among amniotes because, though the end products can be greatly different, embryonic stages are highly similar, especially at the beginning. But genetics alone does not determine the developmental program. The environment in which an embryo develops has significant effects. Development can be affected by factors such as temperature, osmotic conditions, nutrient availability, and so on. The terrestrial environment is an inhospitable place for an egg; it is even more inhospitable for an embryo. Perturbations of the environment can result in embryonic failure. Selection should act to favor adaptations that result in an increased probability that embryos develop within a favorable environment. This does not necessarily mean an invariant environment; for one thing, that would be extremely difficult to find, and for another, variation can be adaptive. Changing conditions could serve as a developmental signal to the embryos. However, retaining the egg within the body offers, at the least, a more predictable and stable environment that results in more predictable and stable embryonic development. Just as there is adaptive value in homeostatic mechanisms that act to buffer the internal environment of an organism from external perturbations, there is adaptive value to mechanisms that buffer the developing embryo. The embryo is unlikely to have the resources or mechanisms in place to be homeostatic; the provision of a more stable, protected environment is, in one sense, the first crucial maternal resource that is provided by retaining the embryo within the reproductive tract. As most adaptations do, it provides challenges as well as advantages.

One challenge is that there must be some kind of barrier between the egg/embryo and the mother to inhibit immunological responses. The innate immune system is ancient. A developing fertilized egg will be seen as "not self" by the maternal immune system. The chorion of placental mammals probably serves an immunological barrier function, to some extent. Thus, although it is not clear that an amniotic egg requires internal fertilization, the amniotic egg certainly has adaptive characteristics that enable internal fertilization and the extended retention of the fertilized egg. And doubtless, extended retention of eggs is a necessity, and probably a required first evolutionary phase, for live birth.

Bony Fishes

There are many species of bony fish that produce live-born young. Live birth would appear to have evolved independently (and been lost) multiple times in this lineage. However, it is not common. Only 14 of the 425 families of bony fish are ovoviviparous or viviparous (Wooding and Burton, 2008).

Bony fish lack a uterus. In live-bearing fish species, the developing embryo resides either in the ovarian follicle or in an ovarian cavity after follicle rupture (Wourms, 1981). About half of the live-bearing fish families have evolved some form of placenta-like structure to enhance maternal transfer of resources to the embryos and are truly viviparous. For the other species, the maternal environment is merely a safe place for the embryo to develop using the materials stored in the yolk (ovoviviparity).

Several different adaptations in viviparous bony fish support live birth. In some species extensions of the embryonic hind gut, called trophotaeniae, come into close apposition, but not permanent contact, with the ovarian epithelium (Lombardi and Wourms, 1985). The ovarian endothelial cells secrete materials that are absorbed by the cells of the trophotaeniae. In three families of bony fish, an extraembryonic sac forms from a coelomic cavity around the heart. This pericardial sac is highly vascularized and comes into close apposition with well-vascularized follicular epithelium, allowing gases, nutrients, and waste products to diffuse across the cell barriers.

The lobe-finned fishes are one of the oldest groups known to have evolved live birth. The coelacanth, a fish often called a living fossil, produces live-born young; however, coelacanths are ovoviviparous. They produce large, tennis-ball-size eggs with a large amount of yolk. These eggs are retained within the female while the embryo grows and develops, using the resources from the yolk. Thus, coelacanths are not truly viviparous; there does not appear to be any further transfer of maternal resources after the egg yolk is formed. Modern coelacanth species tend to be found in deep water, usually several hundred feet down. This might be considered an inhospitable environment for an egg or a developing embryo due to the intense water pressure; coelacanth females provide a protected environment for their offspring by retaining them within their bodies.

Amphibians

Among amphibians, there are few viviparous anurans (frogs and toads) and salamanders, but over a dozen species of apoda (legless burrowers) have evolved viviparity. Most amphibians utilize external fertilization of the eggs, but the apoda practice internal fertilization by intromission.

Caecilians are legless, burrowing amphibians (apoda) with a largely tropical distribution. They are either viviparous or oviparous, with the majority of species studied being viviparous. The embryos of viviparous caecilians develop a specialized fetal dentition that they use to scrape the hypertrophied lining of the maternal oviduct to obtain their nutrition before birth (Wake and Dickie, 1998). Interestingly, some hatchlings of oviparous species are born with similar teeth, which they use to feed on maternal skin. Brooding females of the species *Boulengerula taitanus* produce a lipid-rich, thick skin which their offspring peel off and consume (Kupfer et al., 2006). In these species offspring literally feed on their mothers, sometimes after birth and sometimes before.

An intriguing adaptation for a variant of live birth was found in the gastric brooding frogs, found only in Australia. As the name implies, the female swallows the eggs after they are fertilized. The eggs hatch and the tadpoles develop within the stomach. The eggs and later the tadpoles produce mucus that contains prostaglandins. Prostaglandins have many functions; one is to inhibit gastric-acid secretion. This physiological effect is used in human medicine to treat heartburn and acid reflux disease. Thus the stomach of the female gastric brooding frog becomes a brooding pouch instead of a digestive organ. The female does not eat during the two-week period in which the tadpoles develop, undergo metamorphosis, and then are regurgitated as froglets. The female does not appear to provide nutrients or other chemical signals to the tadpoles, though that is not known for sure. The tadpoles are thought to rely on their large yolk sac for the materials required for growth and development. This adaptation provides a protected and controlled environment, no more and no less. Unfortunately, this fascinating reproductive adaptation likely no longer exists, as the gastric brooding frog was last seen alive in the 1980s and is believed to have become extinct due to human disturbance of their habitat.

Reptiles

Many of the squamate reptiles (lizards and snakes) produce live-born young. The other reptile lineages appear to be precluded from producing live-born young, possibly because of the evolution of a highly mineralized shell. Within squamates, viviparity has evolved independently at least 108 times (Blackburn, 2006). There is considerable diversity in how squamate reptiles have solved the problem of producing live-born young. Although many of the snakes and lizards that produce live-born young practice ovoviviparity, some snakes and lizards develop placentas of varying degrees of complexity. Some skink species have quite complex, well developed chorioallantoic placentas that nourish their young before birth. For example, species of skinks in the genus *Mabuya,* a neotropical group of skinks found in South and Central America and the Caribbean, produce neonates that are hundreds of times more massive than the original fertilized ova (Ramírez-Pinilla et al., 2006), firm evidence for extensive transfer of maternal resources via what is possibly the most complex squamate placenta (Ramírez-Pinilla et al., 2006; Wooding et al., 2010). Immunocytochemistry on placentas from an Andean species showed that transporters for calcium, water, and glucose were localized to specific areas of the placenta, indicating both active transport of nutrients and functional cell differentiation in different placental structures (Wooding et al., 2010). A significant difference between reptilian and mammalian placental function is that in the reptiles, the primary energy substrate transferred from mother to embryo is lipid (Speake et al., 2004), while in mammals it is glucose. In accordance with this difference, the level of expression of glucose transporters in this *Mabuya* species was low, and antibodies to several glucose transporters tested showed no binding at all (Wooding et al., 2010).

Birds

Birds are the only vertebrate class that has no viviparous or even ovoviviparous members. There is no evidence that birds ever evolved live birth. Birds have all of the extraembryonic membranes; they are amniotes. However, birds rely on a massive amount of yolk to provide the needed resources for embryonic growth and development and a well-mineralized shell and a high degree of parental care to protect the embryo after the egg leaves the mother's body.

Feeding Embryos

In all species that produce live-born young, some means of providing the developing embryos with nutrients must exist. In some instances, the means is simply a particularly large egg with a nutrient-rich yolk sac. The embryos are retained within the mother, providing an environment protected from external threats, but there is no additional transfer of maternal resources. However, in many species methods to feed the embryos have arisen. Sharks provide some fascinating examples of the variety of ways in which the nutrient requirements of embryos can be met.

Sharks, along with rays and skates, are members of the vertebrate class Chondrichthyes, or the cartilaginous fishes. This is an especially ancient lineage, first appearing on Earth as many as 450 million years ago. These species have evolved a wide range of reproductive strategies, including a number of mechanisms by which to produce live-born young. Sharks practice internal fertilization; male sharks have a pair of claspers with which they introduce sperm into the female via her cloaca. Thus the strategy of egg retention has been available in the shark lineage for a long time. And many species do retain the eggs. Sharks display the full range of female reproductive possibilities, from species that produce eggs, deposited into the environment to develop and hatch with no further maternal input, to species that have complex placentas that nourish the developing embryos until they are born, with some fascinating alternative strategies lying between these two extremes.

In 1948, a shark researcher was probing the uterus of a late-term pregnant sand tiger shark (*Carcharias taurus*). He was quite startled when he was bitten—the embryos had teeth! In the sand tiger shark, multiple embryos develop from internally fertilized eggs with large yolk sacs. Once they have used up their yolk sac, they begin to feed on each other, a strategy termed *embryophagy*. Eventually, a single offspring is left alive in each of the two uteri. By this somewhat grisly reproductive strategy, a nine-foot-long sand tiger shark can produce two three-foot-long offspring.

In many other sharks, there is a less gruesome form of this strategy called *oophagy*. In other words, the developing embryos feed on eggs. In some spe-

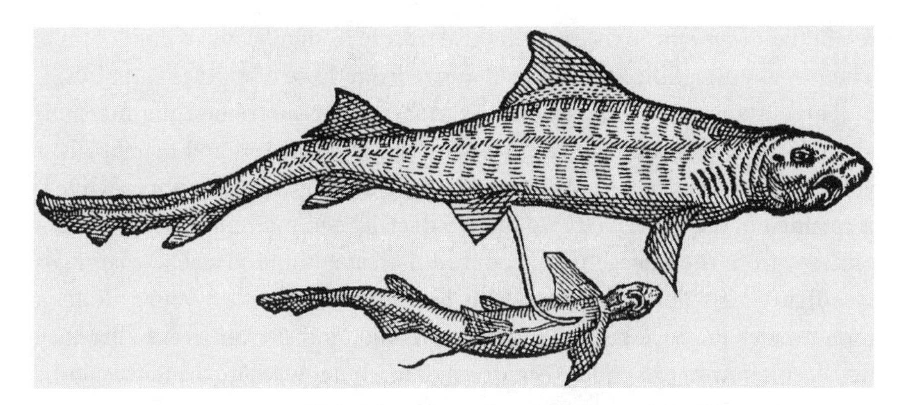

Figure 2.2. Drawing of the dusky smooth hound (*Mustelus canis*) with newly born and still-attached pup, by the French physician G. Rondelet (1507–1556), published in his book *Libri de Piscibus Marinis* and based on descriptions by Aristotle.

cies these eggs may be fertilized, so it is in effect siblicide, just as in the sand tiger shark. However, in some species the mother continues to produce eggs that are not fertilized; in this way she produces a constant stream of nutrition for her embryos.

The final example from sharks is that of true viviparity. In about 27% of shark species, after most of the yolk-sac nutrients are used up, the yolk sac becomes attached to the uterine wall, forming a yolk-sac placenta. This yolk-sac placenta is attached to the embryo by a stem that serves as an umbilical cord. Through this stem nutrients flow from maternal circulation to the embryo, and waste products flow in the reverse direction. In some species the uterine cells secrete fluid called "uterine milk," which can also be absorbed by the yolk-sac placenta and/or the umbilical cord. The first person to describe this phenomenon was Aristotle, from his examinations of the dusky smooth hound (*Mustelus canis*) caught by Mediterranean fishermen in the fourth century BC (figure 2.2).

Mammals and Live Birth

Humans are mammals; as fascinating as the adaptations for live birth in sharks, snakes, and other creatures may be, the best starting place to understand the physiological and disease implications of human adaptations to live birth is mammals. The ancestral mammalian lineage arose hundreds of millions of years ago. There are many kinds of mammals, most extinct. All extant mammals retain their eggs within the reproductive tract for a significant amount of time. These eggs are significantly reduced in size from those of reptiles and birds.

Three mammalian lineages persist today: the monotremes, the marsupials, and the eutherians, or placentals. In both monotremes and marsupials, a yolk-sac placenta develops that nourishes the developing egg/embryo while it is retained in the maternal reproductive tract. The monotremes are an ancient offshoot from the lineage that produced marsupials and placental mammals (see figure 2.1). The monotremes do not produce live-born young. Rather, monotremes produce shelled eggs that are laid, and several weeks later they hatch, either in an egg pouch (echidnas) or in a burrow where the mother curls around them (platypuses). In the two extant monotreme taxa, the embryos develop within the mother for about twice as long as they develop outside of the mother before hatching.

Marsupials produce live-born young; however, gestation lengths are extremely short, on the order of weeks rather than months. Marsupials give birth to embryonic young that crawl to the mother's pouch and attach to the nipples. Marsupial young spend far more time growing and developing during lactation than they do during gestation.

Human beings are placental mammals. As the common name of our lineage suggests, a key adaptation that separates mammals like us from the marsupials and monotremes is a complex, well developed placenta. Monotremes and marsupials do possess rudimentary placentas derived from the yolk sac that nourish first the egg and then, in marsupials, nourish the developing fetus for a short time. There are even a few species of marsupials in which a chorio-allantoic placenta forms shortly before birth. However, most nutrient transfer from mother to offspring in monotreme and marsupial mammals occurs through milk. Indeed, lactation is a defining adaptive characteristic of mammals. In all mammalian taxa, reproduction requires a large transfer of nutrients from mother to offspring, primarily through lactation. However, it can be argued that placental mammals have shifted some of the maternal costs of development from lactation to gestation. In order to do so, a more complex placenta had to develop.

In this book we must give short shrift to marsupial placentas. The marsupials have not been as well studied as the placental mammals, but we know that there is a great deal of variation in the anatomy and complexity of placentas across marsupial taxa. The reader is directed to works by Marilyn Renfree (e.g.,

Renfree and Shaw, 1996; Renfree, 2000; Renfree et al., 2008) for some fascinating studies of marsupial placentas and reproduction.

Mammalian Evolutionary History

In this section we travel far back in time to the early ages of land vertebrates. Life originated in water; initially, land masses contained little life and what was there lived in water. Over time, land was invaded by living things; plants came first, rapidly followed by invertebrates such as insects. Amphibians began to spend time on land, but they still required water for reproduction; their eggs would desiccate if laid on dry land. With the rise of the amniotes (ancestors of reptiles, dinosaurs, birds, and mammals), vertebrates existed that could live their entire life cycle outside of water. The amniotic egg,with its extraembryonic membranes, protects the egg from desiccation and thus enables it to be laid on dry land. These membranes also allow greater transfer of maternal resources to offspring. We begin our review of mammalian evolution with early amniote lineages.

Synapsids and Sauropsids

There are two surviving ancient lineages of amniotes: the synapsids and the sauropsids. Mammals derive from synapsid ancestors; indeed, mammals are the only living descendants of the synapsids. Synapsids are a class of animals that originated more than three hundred million years ago in the Carboniferous era and rose to dominance during the Permian period, between 260 and 286 million years ago (table 2.1). Synapsids are defined by the existence of a single temporal fenestra (hole) behind each eye orbit. Fenestrae serve as attachment points for the jaw muscles and provided adaptive advantages in chewing. In living mammals, the temporal fenestra has become closed by the sphenoid bone, but its developmental origins remain. Synapsids also had differentiated teeth; that is, they had more than one kind of tooth in their mouth. In their living descendants, the mammals, the tooth types are incisors, canines, premolars, and molars. *Dimetrodon,* an apex predator synapsid from the middle Permian, had two kinds of teeth—shearing teeth and sharp canine teeth. Both of these adaptations, jaw muscle attachments and differentiated teeth, are hypothesized to have provided significant adaptive advantages in food processing before digestion and thus to have increased the efficiency and rate at which nutrients could be extracted from food.

Table 2.1 Important biological and geological events

Era	Period	Important Events
Cenozoic	Quaternary (1.8 MYA–today)	Genus *Homo* is well established.
	Tertiary (65–1.8 MYA)	Beginning of rapid radiation of the four placental mammal lineages.
Mesozoic	Cretaceous (146–65 MYA)	Origin of all four placental mammal lineages. Extinction of most dinosaur lineages, except for birds, at the end of the Cretaceous.
	Jurassic (208–146 MYA)	Dominance of the dinosaurs. Pangaea splits into Laurasia and Gondwana. First evidence of hair in the lineage leading to mammals.
	Triassic (251–208 MYA)	Beginning of the dominance of sauropsids, crocodilians, turtles, tuatara, squamate reptiles (lizards and snakes), and dinosaur lineage (birds). Synapsid lineage reduced, with many small, nocturnal species.
Paleozoic	Permian (286–251 MYA)	Age of the synapsids, the ancestors of mammals. Mass extinction at end of Permian.
	Carboniferous (360–286 MYA)	Live birth evolves in placoderms. Extensive colonization of land by vertebrates. The supercontinent Pangaea forms. Amniotes arise.

Note: MYA = million years ago.

The other major amniote radiation that arose at that time was the sauropsids, which gave rise to lizards, snakes, crocodiles, turtles, tuatara, and dinosaurs and their descendants, the birds. Sauropsids came from a lineage of anapsids (no fenestra); interestingly, the surviving lineages of sauropsids are all descended from an offshoot of the anapsids called diapsids, which have two fenestrae on each side of the head. Turtles have an anapsid-like skull, but that is believed to be derived secondarily from a diapsid ancestry.

During the Permian, synapsids occupied a large number of niches. There were large and small synapsids—carnivores and herbivores. Throughout the middle Permian, the most common synapsids were the pelycosaurs, the dominant vertebrates of the Permian ecosystem. The apex predator *Dimetrodon* belonged to a group of pelycosaurs classified as part of a clade called Sphenacodontia. This clade of synapsids gave rise to the therapsids in the late Permian. The therapsids replaced the pelycosaurs as the dominant land vertebrates until the Permian-Triassic extinction event. Mammals are the only living descendants of the therapsids.

At the boundary of the Permian and Triassic periods approximately 250 million years ago (251 ± 0.4 million years ago by the best dating), a worldwide mass extinction event occurred. This mass extinction included the only known mass extinction of insects; at least eight orders of prehistoric insects ceased to exist after this event. It is estimated that up to 70% of the existing land vertebrate species became extinct, including most of the synapsids and sauropsids. The marine ecosystem was hit even harder, with over 90% of species becoming extinct. The cause of the mass extinction is debated; indeed, there likely were multiple reasons. Certainly during this time period there was extensive volcanic activity in what is now Siberia. The Siberian Traps (the word *traps* derives from the Swedish word for "stairs" and refers to the step-like hills of this region) were created about 250 million years ago by massive volcanic eruptions that lasted hundreds of thousands of years. The Siberian eruptions have been dated from 251.7 ± 0.4 million years ago to 251.1 ± 0.3 million years ago (Kamo et al., 2003), remarkably coincident with the Permian-Triassic mass extinction. The result of these prolonged, extensive volcanic eruptions was a massive release into the atmosphere of sulfur dioxide, carbon dioxide from hydrocarbon deposits in that area, and methane, an even more potent greenhouse gas. The huge release of these gases would have driven extensive global warming that increased ocean temperatures. Warmer oceanic waters and higher atmospheric carbon dioxide would probably have resulted in lower oxygen content and higher carbon dioxide content in these waters. Carbon dioxide is about 25 times more soluble in water than is oxygen. The ocean probably became relatively anoxic over much of its extent, though fossil evidence indicates that isolated coastal areas, where wave action may have maintained water oxygen content, appear to have served as refuges for marine life from the extinctions sweeping the world. The anoxic ocean conditions would have tipped the competitive balance in favor of anaerobic sulfur-reducing bacteria, which would have increased water hydrogen-sulfide content. Upwellings of hydrogen-sulfide-laden water would have released the H_2S into the atmosphere. One effect of atmospheric hydrogen sulfide is a degradation of the ozone layer, leading to increased ultraviolet radiation reaching the Earth's surface. It would appear that the world became a relatively inhospitable place for life for a considerable amount of time. But some life survived, the volcanic activity subsided, and during the Triassic the diversity of living things began to increase.

The synapsids had dominated the Permian. During the Triassic it was the sauropsids' turn. The first turtles and crocodilians appeared during this period. The Sphenodontia, a group of lizard-like animals, were widespread. From this lineage only two species (*Sphenodon punctatus* and *S. guntheri*) remain extant, living on islands off New Zealand. These *Sphenodon* species, commonly called *tuataras,* a word derived from the Maori language meaning "peaks on back," retain the primitive diapsid skull morphology. About 230 million years ago, the lineage leading to lizards and snakes diverged from the tuatara lineage (see figure 2.1) and ultimately proved to be more successful than the tuataras. Most significantly, a lineage of sauropsid archosaurs began to expand and radiate. These became the dinosaurs, which eventually became the dominant vertebrate life form from the late Triassic through the Jurassic and up to the end of the Cretaceous.

During the late Permian, a group of synapsids known as cynodonts arose and were eminently successful. Some cynodonts survived the Permian-Triassic mass extinction. Based on aspects of jaw and tooth morphology, mammals were determined to be descended from a lineage of cynodonts. Although cynodonts survived into the Triassic, they were no longer as plentiful as they were in the late Permian. In addition, these descendants were smaller on average than their ancestors. It appears that the surviving synapsids were generally restricted to nocturnal niches. These descendants of large, apex predators had become small, nocturnal insectivores.

Rise of Mammals

This shift to a nocturnal life-style may have been a key event in the evolutionary history of mammals. Some researchers suggest that this specialization in nocturnal living selected for a higher metabolic rate and the ability to regulate body temperature by internal rather than external means. Because of their nocturnal habit, most of the synapsids of the dinosaur era could not raise their body temperature by basking in the sun. Instead, metabolic mechanisms to regulate body temperature evolved. Insulation against thermal losses would also be important. This could be accomplished by behavioral means, for example, by burrowing, nest building, or huddling in a group. Some cynodonts in the early Triassic were burrowing animals. In one fossil bed, multiple individuals were found to have died, possibly from a flash flood, in the same burrow system, implying that these were social animals. Hair or fur would also have a thermal insulation function; as of 164 million years ago, animals with fur existed (Ji et al., 2006).

A nocturnal lifestyle also places selective pressure on sensory systems other than vision. Evolution likely would have favored improved hearing and sense of smell. At this time, the mammalian middle ear developed. Mammals can be defined by the use of two cranial bones for hearing that in the sauropsids are used for feeding. The articular and quadrate bones form the jaw joint in most amniotes, but in mammals, these bones have become the incus and the malleus of the inner ear.

When did true mammals come into being? Initial classification schemes divided mammals into three groups: monotremes were in the prototheria (first animals); marsupials were in the metatheria (changed animals); and mammals were eutheria (true animals). This scheme owed more to human bias than it did to appropriate cladistic classification. The ancient (and continually discredited) idea of a progressive evolutionary change to create more highly evolved lineages (of which we were seen as the peak achievement) has confounded understanding of the evolution of current mammals. If the reader takes anything from this book, let it be that all extant species have had the same amount of evolutionary time over which their lineage has been successful. A successfully evolved species is one that still exists. The number of ways to be successful is so great as to be unimaginable.

Current thought divides extant mammals into two subclasses: the prototheria and the theria. The monotremes still comprise the prototheria, but now marsupials and placental mammals are combined into the theria. The terms *metatheria* and *eutheria* are still used for the lineages leading to marsupials and placental mammals, respectively.

Monotremes

The earliest known fossil monotreme is *Teinolophos,* which lived about 115 million years ago. *Teinolophos* is a member of the platypus group of monotremes (Rich et al., 2001). Living monotremes are represented by a single species of platypus and four species of echidnas. Echidnas are insectivores; the most common species, the short-beaked echidna, eats ants and termites and is widespread in Australia. Platypuses are semi-aquatic, feeding mainly on aquatic invertebrates. Because the two extant taxa are so different, the crown group monotremes (the taxa that all monotremes are descended from) must have evolved well before 115 million years ago. The placental lineage of mammals probably arose less than 110 million years ago, based on molecular data, although this date is later than the dates supported by fossil evidence.

Placental Mammals

Living placental mammals can be distinguished from all other mammals by the fact that they lack the epipubic bones, a pair of bones that extends forward from the pelvic bones in monotremes and marsupials. This is the primitive condition and exists in all extinct nonplacental mammals and in their cynodont ancestors as well. The existence of these bones has been known since 1698, but their function is unresolved. Some have suggested that, at least in the marsupials, they serve to support the marsupium (the pouch) and the joeys within (Tyson, 1698; White, 1989). These bones would be of direct adaptive significance only to females; they would have no function for males beyond an indirect (but nonetheless important) effect of enhancing the reproductive performance of their mates and their own mother and daughters. More recent research has shown that the epipubic bones function to stiffen the trunk during locomotion (Reilly and White, 2003). Reilly and White suggest that epipubic bones were an important locomotor adaptation in the change from a sprawling limb posture common to tetrapod ancestors to a more upright one, with benefits of increased locomotor efficiency. These locomotor benefits would accrue to both sexes.

If the epipubic bones serve an adaptive locomotor function, why did placental mammals lose them? The answer is probably that the existence of epipubic bones would complicate the expansion of the uterus within the pelvic region. The small, embryonic offspring to which marsupials give birth do not require a large expansion of the diaphragm; the larger offspring birthed by placental mammals do. The loss of the epipubic bones would appear to be linked to the increased gestation length in placental mammals and their subsequent larger, more precocial offspring.

When did live birth with a eutherian mammal placenta arise in the mammalian lineage? The fossil record is scanty, but existing fossil evidence and evidence from molecular studies on living taxa suggest that the crown group placental mammals arose more than a hundred million years ago. A purported stem-group placental mammal was discovered in China and dated to about 125 million years ago (Ji et al., 2002). Although that sounds ancient, other characteristics of mammals are older still—for example, hair and, interestingly, lactation preceded live birth and the placenta.

Lactation Preceded Placentation

Lactation is a defining characteristic of extant mammals. In other words, all mammals alive today feed their young with a glandular secretion called milk that is produced by modified apocrine glands in mammary tissue. Even the egg-laying monotremes feed their offspring on glandular secretions of lipid, carbohydrate, and protein (i.e., milk.) Indeed, the term *mammal* derives from the word *mammary*. Based on molecular evidence for the origin of milk protein genes (e.g., caseins and lactalbumin), lactation probably arose more than two hundred million years ago. The casein gene family appears to have originated more than 300 million years ago. Thus lactation most likely arose well before live birth, but possibly not before egg retention.

The original function of lactation is unclear. It has been proposed to have had fluid balance, antimicrobial, or nutritional functions (Oftedal, 2002a, b). At some point, lactation became the primary mammalian adaptation for transfer of maternal resources to offspring. This adaptation enabled a reduction in maternal resources placed into the egg. Specifically, the role of the yolk has decreased greatly over time in the mammalian lineage, evident in the loss of the vitellogenin protein genes in mammals (box 2.1).

Embryo Retention in Mammals

Although monotremes lay eggs, to a large extent embryonic development takes place within the mother. For example, platypus eggs develop in utero for about 28 days before they are laid. The mother then curls around them to incubate them for an additional 10 days before they hatch. In echidnas, eggs are laid about 20–28 days after mating and again take about 10 days to hatch. This contrasts with birds such as chickens, in which the egg develops for only about 24 hours within the female before being laid and is then incubated for 20 or more days until it hatches. In all mammals, a significant amount of embryonic development occurs in utero.

In monotremes and marsupials, the developing embryo is nourished by a chorio-vitteline or a yolk-sac placenta. In the placental mammals a yolk-sac

BOX 2.1 VITELLOGENIN GENES

In species that rely primarily on the egg yolk for the nutrients and other molecules that support the growth and development of the embryo, the vitellogenin genes play a vital role. They serve to transport nutrients and other molecules into the yolk. These genes are ancient. In the common ancestor of amphibians and amniotes (around 350 million years ago), the resources for the developing embryo (protein, lipids, calcium, and phosphorus) were deposited into the yolk of the egg by vitellogenin proteins. There are two ancestor vitellogenin genes, referred to as *VIT1* and *VITanc,* which likely were present in the common ancestor of amphibians and amniotes (Brawand et al., 2008). In the common ancestor to all extant amniotes (reptiles, birds, and mammals) the *VITanc* gene duplicated, producing the ancestor genes to *VIT2* and *VIT3*. Descendants of the three genes *VIT1, VIT2,* and *VIT3* are found in birds today and are fully expressed (Hillier et al., 2004; Brawand et al., 2008). Birds produce an egg rich in yolk and vitellogenin proteins.

Mammals evolved a novel system of transferring nutrients and other bioactive molecules from mother to offspring that differs from that of other amniotes. They rely on the chorio-allantoic placenta first, and then on milk to provide the maternal resources necessary for the young to grow and develop. Vitollegenin proteins transport calcium and phosphorus, among other nutrients, into egg yolk. Lactation most likely preceded live birth in mammals, probably by more than a hundred million years. The casein proteins in milk efficiently transport calcium and phosphorus. Early mammals relied more on lactation to transfer maternal resources to their offspring; mammalian eggs are small, with little yolk. The vitellogenin proteins were not essential and appear to have been lost.

Pseudogenic remnants of vitollegenin genes (mainly from *VIT1* and *VIT3*) were found in both dog and human genomes in regions syntenic to those containing the bird *VIT* genes. Exon 3 of the pseudogene *VIT1* found in armadillos, dogs, and humans share two indels, implying that *VIT1* became inactivated in a common ancestor of these species. Considering that these species represent three of the four superorders of mammals, *VIT1* was likely not expressed in the stem eutherian mammal. In the marsupial gray short-tailed opossum, coding sequence remnants were found for all three bird *VIT* genes. This implies that these genes date back to the common ancestor of birds and mammals (Brawand et al., 2008). They remain expressed in birds and are vital adaptations for avian reproduction, but they have been silenced in mammals.

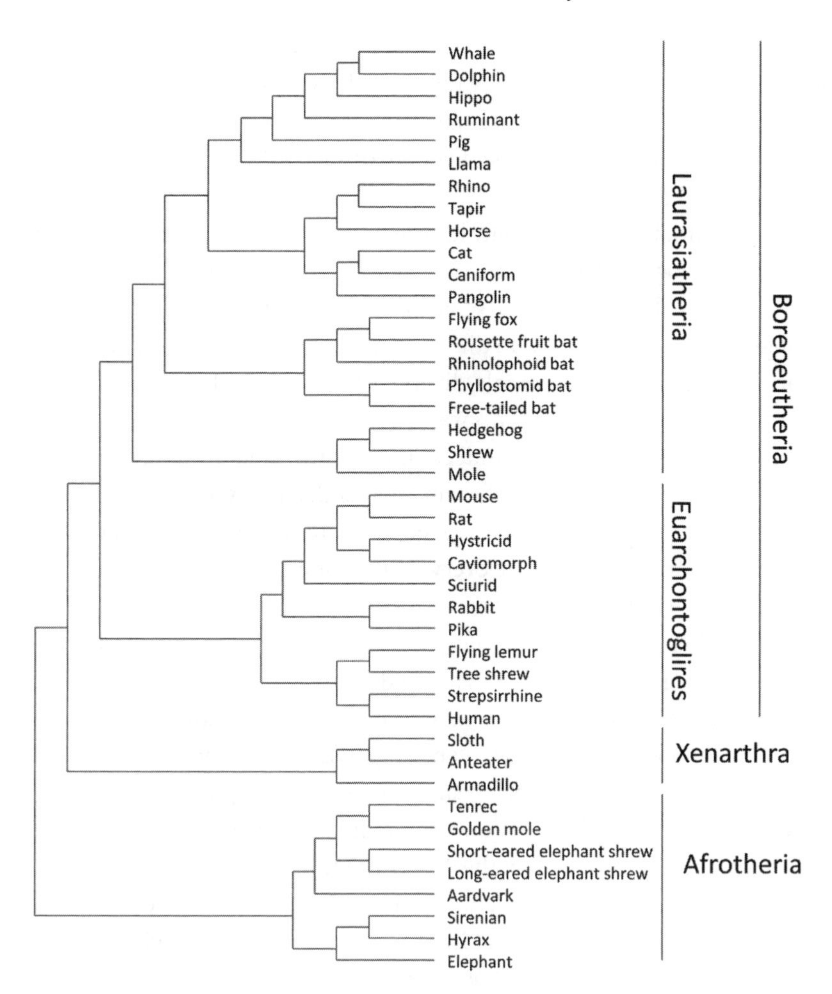

Figure 2.3. Molecular phylogeny of mammals, showing the four superorders.

placenta forms first, but then a true chorio-allantoic placenta forms. The initial yolk-sac placenta plays a variable role in embryonic growth and development in placental mammals, depending on the species.

Eutherian Mammals

The eutherian mammals appear to have diverged from the marsupials roughly 140 million years ago. Eutherian mammals consist of the extant placental mammals and a number of extinct lineages. Placental mammals have been quite successful. There are 18 orders of extant placental mammals (more or less, depending on the taxonomic authority referenced), with a wide diversity of form and life history. A diversity of placental structure exists as well (see table 3.1 in chapter 3), which likely reflects the diversity of successful adaptive planes that various mammal species occupy.

Advances in molecular phylogeny have greatly revised earlier ideas of placental mammal phylogeny. The evidence is strong that there are four superorders of eutherian mammals—Afrotheria, Xenarthra, Laurasiatheria, and Euarchontoglires (figure 2.3)—that evolved in different geographic areas. The existing four superorders of placental mammals diverged from each other approximately one hundred million years ago. This was well before the end of the dinosaurs at the Cretaceous extinction event. Indeed, most of the orders of mammals were in existence before the sequence of events that ended the Age of Dinosaurs. Of course, since that time the different placental mammal orders have undergone an expansive radiation to occupy every continent and to dominate the niche of large terrestrial animals.

Primates belong within Euarchontoglires, along with tree shrews, dermopterans (herbivorous gliding animals often called flying lemurs), rodents, and lagomorphs (rabbits and hares). Laurasiatheria is composed of the cetartiodactyls (e.g., whales, hippos, cows, pigs, and llamas), perissodactyls (horses, rhinos, and tapirs), carnivores (e.g., dogs, cats, bears, and seals), chiroptera (bats), and the group composed of hedgehogs, shrews, and moles. The Afrotheria encompasses animals that originally evolved on the African continent and includes elephants, hyraxes, golden moles, and manatees. Species in Xenarthra include armadillos, sloths, and anteaters, all species that live in the Americas. Originally these taxa were grouped with pangolins and aardvarks in the order Edentata, so named because all these species either lack incisor teeth and molar teeth or have poorly developed teeth. The molecular evidence showed that Edentata was paraphyletic (contained more than one lineage of organisms). Pangolins and aardvarks are now classed in Laurasiatheria and Afrotheria, respectively, in completely different superorders from their former Edentata companions.

Now that a well-established phylogeny of the placental mammals exists, a comparative examination of placental types across species can be fruitfully undertaken to consider questions such as, what did the placenta of the ancestral placental mammal look like? There is, as yet, no definitive answer to this question, but the distribution of placental morphology across the mammalian phylogeny indicates that the placenta may be the most changeable of organs.

3

Comparative Mammalian Placentation

The placental mammal radiation has resulted in a wide range of species in terms of ecology, life history, and development. Not surprisingly, this diversity is reflected in a large variation in the level of neonatal development at birth among mammalian species. Neonates can be altricial (poorly developed) or precocial (well developed). For example, horse foals and giraffe calves are capable of competent locomotion within hours of birth (highly precocial), while many carnivore neonates are born blind and effectively helpless (altricial). Human neonates fall in between; like most anthropoid primates, they have precocial characteristics (eyes and ears open and functional). However, they cannot locomote independently for a considerable time after birth. Human neonates have been suggested to be secondarily altricial relative to other anthropoids, due to their relative helplessness at birth and their extended infancy period. Of course, all mammals share a requirement for maternal postpartum care, as the first food of all mammals is mother's milk. No mammalian neonates are fully independent at birth, though there are some marine mammal species in which postpartum care is short and restricted to a rapid and substantial transfer of nutrients via milk. But as a rule, mammals can be characterized by extensive prenatal maternal effort, followed by an even more extensive postpartum investment of maternal time and resources.

Given the diversity among placental mammals that exists at all life stages, it is not surprising that placental morphology and development differ widely among taxa. In some cases, it seems apparent that different species are solving different challenges, which require diverse solutions. In other instances, it appears that different species are solving similar challenges in dissimilar ways. In biology, there is rarely a single path to success. All placental forms that are found in extant species represent, by definition, successful adaptations.

The essential function of a placenta, in the evolutionary sense, is to produce a neonate that will be capable of the appropriate growth and development necessary to become a viable adult without significantly degrading the maternal capacity to provide the necessary postpartum care. This is an important constraint, considering that lactation is an essential component of all mammalian reproduction, and that lactation generally produces a larger transfer of maternal resources to offspring than occurs during gestation. In addition, it can be argued that the capacity for the mother to produce future offspring should be minimally compromised, though that is a potential area of maternal-fetal and maternal-paternal conflict.

The modern mammalian placenta is a fascinating transient organ with multiple functions. It connects two genetically distinct organisms and links their physiology and metabolism. It certainly functions to pass required nutrients and gases to the fetus; it is the conduit for the molecules of life necessary for fetal metabolism. However, the placenta has regulatory and immunological functions as well. The placenta develops from the fertilized egg, serves key functions that allow the fetus to develop, and then separates from the uterus and is expelled after birth. It is primarily composed of embryonic tissue, usually referred to as extraembryonic, because it in effect becomes a separate entity from the embryo, though it is genetically identical. It is derived from the amniotic membranes (amnion, chorion, and allantois) of the fertilized ovum and therefore represents an adaptation of an ancient structure for new function.

In this chapter we explore this diversity of placental development, form, and cellular morphology and discuss the possible functional consequences of the different placental types. We consider a set of questions regarding the evolution of the placenta: what advantages and constraints do the different placental types convey, and what might have been the original placental form in the common ancestor of the placental mammals?

Interhemal Distance

The specific functions of the placenta that enable it to perform its evolved purpose mostly involve regulating the transport of molecules between the maternal and fetal blood circulations. Nutrients and oxygen need to go from maternal to fetal circulation; carbon dioxide, urea, and other waste products of metabolism need to go from fetal to maternal circulation. But many other molecules besides nutrients and wastes need to be transported. For example, in anthropoid primates and horses, chorionic gonadotropin (CG) is produced by the placenta and secreted into maternal circulation. The ovaries are stimulated by CG to produce progesterone, which is required to maintain the uterus in a receptive condition; thus the placenta is sending a signal to the maternal ovaries that a viable conceptus has implanted. If there is no CG signal, ovarian progesterone production will rapidly decline, and the uterus effectively will reject the conceptus and terminate the pregnancy.

A mammalian placenta contains two blood streams within it. About half the volume of a human placenta is blood channels. Maternal and fetal circulation are always kept separate (though there can be some transfer of cells back and forth). A tissue barrier always exists between the maternal and fetal blood. Molecules cross the placenta either by diffusion or some form of active or facilitated transport. In the case of diffusion, the ability for molecules to cross the placenta in either direction is strongly influenced by the interhemal distance, or the thickness of the cellular barrier between the maternal and fetal blood. The interhemal distance is not the only constraint; many other parameters can affect the efficiency of transfer between the maternal and fetal circulation, especially if active transport is involved. However, a small interhemal distance generally will increase the rate at which molecules can transfer between maternal and fetal blood, either by diffusion or active transport.

The interhemal distance can be reduced by thinning the layers of cells between the maternal and fetal blood vessels or by reducing the number of layers. Both methods can be equally effective. For example, in the pig placenta (epitheliochorial) all six layers of cells between maternal and fetal blood (three fetal and three maternal) remain. In the human placenta (hemomonochorial) there are only three layers; the three maternal layers have been degraded during the implantation process and only the three fetal cell layers remain. It is true that the mean interhemal distance in the pig placenta is greater than the mean in the human placenta; fewer cell layers translate to a smaller interhemal distance, on average. However, the minimal distance is about the same. There are regions of the pig placenta where the six cell layers have become sufficiently thinned to equal the minimal interhemal distance of the three cell layers in a human placenta.

The interhemal distance varies among species but can also vary substantially in different regions of a placenta. There is evidence that in the human placenta, the interhemal distance is associated with oxygen concentration; areas of minimal interhemal distance are more extensive in placental regions where oxygen concentration is probably low, presumably to facilitate oxygen diffusion. Thus the interhemal distance can be flexible, and the placenta can react to differences in nutrient concentrations locally as well as globally.

Placental Tissue

Of what kinds of tissues are placentas made? The placenta is an organ, just as the liver, kidney, heart, or skin is an organ. It is made up of specialized (and some nonspecialized) cells that perform the morphological, physiological, and metabolic functions that comprise the various aspects of placental functions. The extraembryonic membranes that make up the placenta are the yolk sac, the chorion, the amnion, and the allantois. All species begin gestation with a yolk-sac placenta; for some species the yolk sac persists for a considerable time. Eventually, however, all placental mammals produce a chorio-allantoic placenta, in which the allantois fuses with the chorion and the amnion forms a fluid-filled sac surrounding the fetus.

The outermost layer of the placenta is composed of trophoblast or tissue derived from trophoblast. This outermost placental tissue is epithelium with a high capacity for cell fusion, hormone production, active metabolism, nutrient transport, and resistance to maternal immunological attack. Cytotrophoblasts are mononuclear trophoblast cells from which other cell types arise.

Trophoblast Giant Cells

Trophoblast giant cells (TGCs) are characteristic of rodent placentas. They are mononuclear but usually polyploid, phagocytic, and invasive. They are generally large (50–100 μm in diameter), but not always. The nucleus is polytenic, with each chromosome having multiple copies that remain attached. They play an important role in rodent implantation by eroding maternal epithelium and thereby enlarging the implantation chamber. Although TGCs are a feature of rodent placentas, their function and gene expression are analogous (and perhaps in some cases homologous) to extravillous cytotrophobast cells in human placentas. Both of these cell types are invasive and play important roles in eroding and remodeling maternal tissue. Many genes involved in the development and function of TGCs are conserved between rodents and humans (Hu and Cross, 2010).

There are four main types of TGCs in rodent placentas: parietal TGCs that line the implantation site and are in direct contact with maternal tissue, spiral artery–associated TGCs; maternal blood canal–associated TGCs; and sinusoidal TGCs within the sinusoidal blood spaces of the labyrinth. They can be distinguished not just by anatomical location but also by function and gene expression. All are secretory cells, their major function associated with secreted molecules that affect adhesion of the conceptus to the uterus, erosion and remodeling of maternal tissue, and regulation of maternal metabolism. Importantly, TGCs are not proliferative (Hu and Cross, 2010). They undergo DNA duplication without mitosis, in a process termed endoreduplication, which enables them to accumulate large amounts of DNA within the nucleus. Instead of being diploid (chromosome count = 2N) they can have a chromosome count of up to 1,000. Their great size, with associated large amounts of endoplasmic reticulum and multiple DNA copies, may act to increase their protein synthesis and hence their secretory ability (Hu and Cross, 2010). Their large size also allows rapid tissue growth. They are the first placental cells to become terminally differentiated in rodent placentas, and they are important in the formation of the initial yolk-sac placenta. TGCs continue to have important functions that change throughout gestation. These varying functions of TGCs depend on gestation stage as well as anatomical location.

TGCs become functional at the blastocyst stage. During implantation, TGCs secrete progesterone, which primes the uterine epithelial cells to be receptive to implantation; they also secrete various adhesion molecules that enable the blastocyst to attach to the uterine wall. After attachment, TGCs secrete signaling molecules (e.g. progesterone, integrins) that stimulate decidualization of the uterine epithelium. Spiral artery–associated TGCs secrete bioactive molecules that initiate and regulate spiral-artery remodeling (Hu and Cross, 2010). The fact that TGCs are not proliferative may be an essential aspect of their function; TGCs are invasive and can promote local angiogenesis and blood vessel remodeling and thus would have a great potential to cause tumors if they had not left the mitotic cycle (Hu and Cross, 2010).

Trophoblast giant cells are also found in ruminant and human placentas, but they differ from rodent TGCs. Both ruminant and human TGCs are multinucleate. Human TGCs appear to be formed by fusion of cytotrophoblasts, not by endoreduplication. Bovine TCGs are binucleate and polyploid (Klisch et al., 2006), though their nuclear content is not as high as that in rodent TGCs. Bovine TGCs are secretory in function; they fuse with maternal epithelial cells to form trinucleate, hybrid cells (Wooding, 1992) that are able to release cell products (hormones, cytokines, and so forth) into uterine epithelium. Thus they serve as the conduit for placental-fetal information molecules to enter the maternal circulation, bypassing the normally impermeable barrier of the ruminant synepitheliochorial placenta (Klisch et al., 2006).

The Syncytiotrophoblast

Most placental cells are mononuclear (containing only one nucleus), as are most cells in any vertebrate tissue. There are binucleate and multinucleate placental cells as well, however, such as the trophoblast giant cells of rodents, humans, and ruminants, mentioned above. Perhaps the most unusual tissue that occurs in some, but not all, placentas is the multinuclear syncytiotrophoblast.

A syncytium is a cytoplasm-filled, multinucleated, cell-like structure. Perhaps the simplest example is when cells replicate their nuclei but do not undergo cytokinesis. This process results in cells containing two or more nuclei. Cells can also fuse, resulting in two or more nuclei within the cytoplasm. There are many examples of syncytia in nature; multinucleated cells are common in fungi and algae. They are also an important component of vertebrate skeletal muscle fibers. Muscle fibers are formed by the fusion of myoblasts, the mononuclear progenitor cells of skeletal muscle cells. Thus, muscle fibers are composed of multinucleated cells (syncytia) that are formed by cell fusion.

Many, but not all, mammalian placentas have an outer syncytial layer consisting of a multinucleated tissue formed by the fusion of cytotrophoblast cells. This outer layer of tissue is called the syncytiotrophoblast. It has membranes on only the apical (maternal) and basal (fetal) sides. It is tissue that is not composed of cells; rather, it was formed by cells, but the membranes between the cells are gone. The villous trees of the human placenta (the fetal blood vessels that receive maternal resources) are covered by the syncytiotrophoblast. Thus, in humans, the syncytiotrophoblast is the placental tissue that comes into direct contact with maternal blood for most of pregnancy. The syncytiotrophoblast is the primary tissue involved in regulation of maternal immune response and in expression of proteins, hormones, cytokines, chemokines, and all the other information molecules by which the placenta communicates with, and regulates, maternal physiology.

Syncytiotrophoblasts are commonly found in hemochorial and endotheliochorial placentas but rarely, if ever, in epitheliochorial placentas. Species whose outer layer of placental tissue is composed of syncytiotrophoblast include all the anthropoid primates, rodents, and most carnivores. Species that do not have a syncytiotrophoblast include the strepsirrhine primates, the elephants, and the cetartiodactyls; ruminants display an unusual trinucleate, hybrid outer tissue, as mentioned above.

How does a syncytiotrophoblast form? Cytotrophoblast cells fuse due to the action of one or more proteins expressed by cytotrophoblasts when they reach their final stage of differentiation. The fusogenic proteins in humans are called syncytin 1 and syncytin 2. Analogous proteins, syncytin A and syncytin B, have been found in murid rodents (rats and mice). These proteins derive from retroviral envelope protein genes (Dupressoir et al., 2005). They are the co-opted remnants of an ancient retroviral infection in the ancestral anthropoid primate and murid rodent lineages. Another retroviral-derived syncytin protein has been found in rabbits (Heidmann et al., 2009). Interestingly, although the syncytin genes in these different lineages appear somewhat related to each other, they are not orthologs. In other words, they are not related by common descent. The retroviral invasion of the germ line did not occur in the common ancestor of rats, rabbits, and monkeys; rather, multiple independent retroviral infections became established in these different lineages.

The process by which cytotrophoblast cells fuse to form a syncytiotrophoblast is still not completely understood. We know that syncytin 1 and 2 have a role in humans, just as syncytin A and B do in murid rodents. Prior to fusion with the syncytiotrophoblast, the cytotrophoblast cells have become highly differentiated. The nuclei within a syncytiotrophoblast are terminally differentiated—in other words, they have left the cell cycle and are therefore not capable of mitosis. Thus, the number of nuclei in a syncytiotrophoblast can only increase through cytotrophoblast cell fusion. But, not all nuclei within the syncytium are the same. Nuclei within a syncytiotrophoblast age; young nuclei, from recently fused cytotrophoblasts, are more euchromatic and contain a nucleolus. Older nuclei display characteristics similar to nuclei of cells nearing apoptosis, such as increasing chromatin aggregations, and they tend to congregate in what are termed syncytial knots (Huppertz, 2010). These nuclei are eventually sloughed off and enter maternal circulation, where they are engulfed by maternal macrophages in the maternal lungs. Transcription of DNA within syncytiotrophoblast nuclei is downregulated—only some nuclei are producing RNA transcripts. However, the placenta contains proteins that stabilize RNA and block its degradation. For example, milligram quantities of a soluble ribonuclease inhibitor can be extracted from human placentas (Blackburn et al., 1977). Thus, the half-life of RNA in syncytiotrophoblast nuclei is likely considerably longer than in other tissue, possibly allowing protein synthesis at higher levels than would be predicted from RNA transcription.

The nuclei contained in syncytiotrophoblasts produce the peptides that are secreted into maternal circulation. Placental signaling is regulated; the types and amounts of molecules secreted into maternal circulation vary over gestation, sometimes in ways that suggest clock-like regulation. For example, during human gestation chorionic gonadotropin (CG) is secreted by the placenta in large amounts in very early pregnancy and then is generally absent by the second trimester (Jaffe et al., 1969). Corticotropin-releasing hormone secreted by the human placenta is undetectable in maternal circulation until the second trimester and then increases exponentially until parturition (Power and Schulkin, 2006). Gene expression in the nuclei of the syncytiotrophoblast would appear to change over gestation, but how this gene expression is regulated is not yet understood.

Much of the regulation of placental signaling likely derives from feedback loops between maternal and fetal metabolism. Molecules pass from the mother and the fetus to the placenta, affecting its metabolism and ultimately gene expression. Certainly maternal signaling molecules can easily reach the syncytiotrophoblast; fetal signaling molecules undoubtedly can as well. However, the unique structure and origin of the syncytiotrophoblast suggest that another clock-like mechanism to regulate the signaling patterns over gestation may operate.

Syncytiotrophoblasts contain nuclei of different ages; these nuclei came from cytotrophoblasts that differentiated at various times during gestation. Either of these factors—age and differentiation time—can influence gene expression by nuclei. As mentioned above, as nuclei in syncytiotrophoblasts age, they change in ways that are consistent with the assumption that their gene expression patterns change. In general, most gene expression probably decreases, but it is possible that some genes are upregulated, especially if repressor genes are silenced.

All new nuclei come from terminally differentiated cytotrophoblasts. When cytotrophoblast stem cells undergo mitosis, one daughter cell becomes a new stem cell and the other fuses with the syncytiotrophoblast. Thus there is a constant supply of cytotrophoblast stem cells to add to the syncytiotrophoblast, and since syncytiotrophoblast nuclei do not divide, the only way to increase syncytiotrophoblast mass is for differentiated cytotrophoblasts to continually fuse. However, the biochemical environment in which a cytotrophoblast cell differentiates and eventually fuses with the syncytiotrophoblast varies over gestation. In theory, that could affect the patterns of gene expression by the nuclei. Thus, as gestation progresses, the population of nuclei in the syncytiotrophoblast changes, with different ratios of young and old nuclei and nuclei that were created at different time points under different conditions. We speculate that these might be two (of many) mechanisms that operate to change placental signaling over gestation. For example, it would be interesting to see whether particular gene products are localized to similar-aged nuclei and whether those patterns change over gestation.

A syncytial layer is common in hemochorial and endotheliochorial placentas, but it has not been found in species with epitheliochorial placentas to date. There are species with endotheliochorial or hemochorial placentas having an outer cellular layer and no syncytiotrophoblast (see table 3.1). The advantage of a syncytial layer is that it provides a more complete barrier between maternal and fetal blood, as there will be no paracellular transport across the syncytium. A hemomonosyncytiochorial placenta (a single placental layer of syncytiotrophoblast and erosion of all three maternal layers) provides an exceedingly small interhemal distance between maternal and fetal blood that enhances molecular exchange yet at the same time provides a junction-free barrier conducive to the regulation of molecular exchange. Molecules must cross the membrane to flow between the mother and the fetus.

The fact that, in at least two orders of mammals, the placental syncytiotrophoblast appears to require genes from ancient, suppressed retroviruses raises some interesting questions. Does the presence of a synctial tissue layer in a placenta imply the existence of co-opted retroviral genes? To what extent was the placental mammal radiation dependent on ancient retroviral infections? The answers are not clear. Certainly, placental evolution in rodents, lagomorphs, and anthropoid primates appears to have been greatly influenced by one or more ancient retroviral infections that entered the germ line and were then largely silenced by genetic defensive mechanisms, leaving only a few expressed genes that have been incorporated into the species phenotype as well as genotype. But is that the only path to a placenta with a syncytium? There are other examples of syncytia in mammalian tissue, so the answer to that question must be no. However, the genetics of the syncytiotrophoblast has not been investigated in all species that have a placental syncytial layer. Many intriguing questions are yet to be answered. What did the anthropoid primate ancestral placenta look like before the retrovirus was incorporated into the genome of the anthropoid common ancestor? Did it have a syncytial layer?

Characterizing Placentas

Monotremes have a yolk-sac placenta and an allantois that connects the fertilized egg to the uterus. Monotreme eggs are much larger than marsupial or eutherian eggs, about 3 mm in diameter at ovulation. They grow to five to six times that diameter before they are laid, absorbing a large amount of nutrients from maternal tissue through their extraembryonic membranes. This is termed matrophic nutrition, where nutrients are deposited into the egg as it is retained within the reproductive tract. A yolk-sac placenta is important early in gestation for placental mammals but is generally not important or is restricted to specific functions in later gestation. In rodents and lagomorphs, the inverted yolk sac is the main source of transferred immunoglobulins (IgGs). A general finding is that larger molecules can better transfer across a yolk-sac placenta compared to a chorioallantoic placenta. A yolk-sac placenta must meet embryonic needs until the allantois joins with the chorion. But the early gestational period is one more of development than of growth; there is a need for key molecular signals, but the total amount of nutrients is small compared to later in gestation. Initially placental growth, almost by necessity, exceeds fetal growth.

Marsupial mammals have a placenta; in most cases it is formed by a yolk-sac placenta and becomes a chorio-vitteline placenta. However, some species (e.g., bandicoots) develop a chorio-allantoic placenta.The bandicoot may be said to have one of the most invasive placentas; fetal and maternal tissues actually combine to form a synctiotrophoblast (Padykula and Taylor, 1977). Thus fetal and maternal nuclei are next to each other, sharing a cell. Fusion of trophoblast and uterine epithelial cells is not so rare; as mentioned previously, it also occurs in ruminants. Marsupial placentas are not long-lasting—gestation is measured in terms of a few weeks. Lactation is the principle mechanism for maternal transfer of resources to offspring.

As mentioned in the introduction, there are many ways to classify placental structure. We start with classification regarding the number of cell layers between maternal blood and placental tissue. There are three layers of maternal cells between the chorion and maternal blood: uterine epithelial cells, connective tissue, and endothelial cells lining the maternal blood vessels. All placentas have some invasive quality; the placenta is never free floating in the uterus but is always attached to the uterine wall in some manner. With an epitheliochorial placenta there is minimal invasion of maternal tissue, with all three maternal cell layers intact. With an endotheliochorial placenta, trophoblast cells erode the uterine epithelium and connective tissue and invade the maternal endothelium. With a hemochorial placenta we see complete invasion of all three maternal cell layers such that placental trophoblasts are bathed in maternal blood. Another way of expressing these relationships is that with an epitheliochorial placenta, the chorion is in contact with uterine epithelium; with an endotheliochorial placenta, the chorion is in contact with endothelial cells lining maternal blood vessels; and with a hemochorial placenta, the chorion is in direct contact with maternal blood.

In the following sections the four major categories of placentas found in the placental mammals are discussed, in general following the original classification of Grosser (1927), modified by more recent investigators. A list of species with their placental types, organized by mammalian superorder, is presented in table 3.1. The placental types are described generally, but placentas of certain representative species of particular interest and placentas that have been particularly well studied are described in more detail. For more detailed descriptions of the anatomy and cellular structure of these basic types of mammalian placentas, the books by Benirschke and colleagues (2006) and Wooding and Burton (2008) are recommended.

Epitheliochorial Placentas

The basic categorization of placental types starts with the number of layers of cells between maternal and fetal blood. The epitheliochorial placenta retains all cellular layers between maternal and fetal blood vessels, with no erosion or destruction of maternal cells. Trophoblasts are in apposition with uterine epithelium and do not invade. Fetal and maternal microvilli often interdigitate, creating a substantial surface area in close contact. The cell layers may be greatly thinned, despite the lack of any loss of cellular layers, resulting in extremely small diffusion distances. In the pig the interhemal distance can be as little as 2 μm. As pregnancy progresses, fetal and maternal blood vessels may make deep indentations into the uterine and placental tissues, respectively, but there is no uterine decidualization. A syncytium does not form, although experimental implantation of pig blastocysts below the uterine epithelium results in syncytium formation by trophoblasts (Samuel and Perry, 1972). So, although at least some capacity for cellular fusion exists, in normal pig and horse placental development the trophoblast cells appear uniformly uninucleate (Wooding and Burton, 2008). The best-studied epitheliochorial placentas

Table 3.1 Placental morphology in various species from the four major clades of eutherian mammals

Species	Placental Form	Maternal Cell Layers	Trophoblast Layer(s)	Interdigitation	Decidual	Superorder
Elephant	Zonary	Endotheliochorial	Syncytial	Labyrinthine	No	Afrotheria
Rock hyrax	Zonary	Hemochorial	Cellular	Villous	Yes	Afrotheria
Aardvark	Zonary	Endotheliochorial	Cellular	Labyrinthine	Yes	Afrotheria
Manatee	Zonary	Endotheliochorial		Labyrinthine		Afrotheria
Giant elephant shrew	Discoid	Hemochorial	Cellular	Labyrinthine	Yes	Afrotheria
Tenrec	Discoid	Hemochorial	Cellular	Labyrinthine	No	Afrotheria
Giant anteater	Discoid	Hemochorial	Syncytial, cellular	Villous	Yes	Xenarthra
Sloths	Cotyledonary	Endotheliochorial	Syncytial	Labyrinthine	Yes	Xenarthra
Nine-banded armadillo	Zonary, cotyledonary	Hemochorial	Syncytial, cellular	Villous	No	Xenarthra
Hyena	Zonary	Hemochorial	Syncytial, cellular	Villous	No	Laurasiatheria
Raccoon	Zonary	Endotheliochorial	Syncytial, cellular	Labyrinthine	Yes	Laurasiatheria
Red panda	Discoid	Endotheliochorial	Syncytial	Labyrinthine	Yes	Laurasiatheria
Dog	Zonary	Endotheliochorial	Syncytial, cellular	Labyrinthine	No	Laurasiatheria
Cat	Zonary	Endotheliochorial	Syncytial, cellular	Labyrinthine	No	Laurasiatheria
Dama gazelle	Cotyledonary	Epitheliochorial	Cellular	Villous	No	Laurasiatheria
Sheep	Cotyledonary	Epitheliochorial	Syncytial	Villous	No	Laurasiatheria
Cow	Cotyledonary	Epitheliochorial	Cellular	Villous	No	Laurasiatheria

Hippopotamus	Diffuse	Epitheliochorial	Cellular	Villous	No	Laurasiatheria
Warthog	Diffuse	Epitheliochorial	Cellular	Villous	No	Laurasiatheria
Dolphin	Diffuse	Epitheliochorial	Cellular	Villous	No	Laurasiatheria
Vampire bat	Discoid	Hemochorial	Syncytial, cellular	Labyrinthine	No	Laurasiatheria
Rabbit	—	Hemochorial	Syncytial, cellular	Labyrinthine	Yes	Euarchontaglires
Mouse	Discoid	Hemochorial	Syncytial, cellular	Labyrinthine	Yes	Euarchontaglires
Rat	Discoid	Hemochorial	Syncytial, cellular	Labyrinthine	Yes	Euarchontaglires
Guinea pig	Discoid	Hemochorial	Syncytial	Labyrinthine	Yes	Euarchontaglires
Porcupine	Discoid	Hemochorial	Syncytial, cellular	Labyrinthine	Yes	Euarchontaglires
Tree shrew	—	Endotheliochorial	Cellular	Labyrinthine	Yes	Euarchontaglires
Colugo	Discoid	Hemochorial	Syncytial, cellular	—	—	Euarchontaglires
Ruffed lemur	Diffuse	Epitheliochorial	Cellular	Villous	No	Euarchontaglires
Slow loris	Diffuse	Epitheliochorial	Cellular	Villous	No	Euarchontaglires
Galago	Diffuse	Epitheliochorial	Cellular	Folded	No	Euarchontaglires
Tarsier	Discoid	Hemochorial	—	Labyrinthine	Yes	Euarchontaglires
Common marmoset	Discoid	Hemochorial	Syncytial	Trabecular	Yes	Euarchontaglires
Rhesus macaque	Bi-discoid	Hemochorial	Syncytial	Villous	Yes	Euarchontaglires
Human	Discoid	Hemochorial	Syncytial	Villous	Yes	Euarchontaglires

are from pigs and horses. Detailed descriptions of implantation and placental development in these species can be found in Wooding and Burton (2008).

In species with an epitheliochorial placenta, the uterine glands secrete copious amounts of materials that are absorbed, first by the yolk-sac placenta and later by the chorio-allantoic placenta. The placenta forms regions called areolae over the uterine gland mouths. The areolae transport uterine glandular secretions to the fetus. This is termed histiotrophic fetal nutrition, as opposed to hemotrophic fetal nutrition, which refers to nutrients transferred directly from maternal blood. Histiotrophic maternal-fetal transfer is common in non-mammalian placentas and operates to a greater or lesser extent in all mammalian placentas. It is the dominant and often sole means of transporting maternal resources to the placenta-fetus in very early gestation, including early human gestation (Burton et al., 2002).

In species with a diffuse, epitheliochorial placenta, such as pigs, nutrient uptake is generally considered to be a function of placenta size, which determines the amount of surface area in contact with the uterine epithelium (Leiser and Dantzer, 1988). For example, in Yorkshire pigs, placental surface area doubles between days 90 and 110 of gestation, allowing the increase in nutrient transfer necessary at the end of gestation (Biensen et al., 1998). However, in Meishan pigs, placental size does not increase at the end of gestation; instead, placental and endometrial vascularization increases greatly (Biensen et al., 1998).

Species with an epitheliochorial placenta include many cetartiodactyls (e.g., pigs, dolphins), an insectivore (the American mole, but no other insectivores studied so far), and the strepsirrhine primates (the lemurs of Madagascar and the lorises of Asia and Africa). Thus, species with epitheliochorial placentas are representative of multiple mammalian superorders, though predominantly the Laurasiatheria. There is no clear phylogenetic pattern, but the apparent lack of species with epithliochorial placentas in the Xenarthra and Afrotheria would imply that the epitheliochorial placenta type is unlikely to have been the ancestral one.

Many species with epitheliochorial placentas are large and have relatively long gestations. A suggested advantage of an epitheliochorial placenta is that the full set of cell layers between maternal and fetal blood reduces the antigenic risk factor and thus allows for long gestations. This idea is actually more an argument that the immunological challenge should be greater for hemochorial placentas, but the anthropoid primates with generally long gestations and hemochorial placentas indicate that this challenge can and has been solved. And, of course, the American mole and many strepsirrhine primates are small-bodied animals. In fact, the strepsirrhines include both the smallest primate species and also, within recent prehistory, the largest, as the subfossil evidence from Madagascar shows that lemurian primates larger than modern gorillas lived on that island until shortly after human beings arrived (Karanth et al., 2005).

The primary focus of this book is the human placenta; however, a central goal of the book is to encourage broader, more comparative investigations and hypotheses regarding human placental function and adaptations. One aspect of the human placenta that has engendered much academic discussion is its invasive qualities, with a great deal of speculation regarding the adaptive function of this deep invasion of maternal tissue. In general, epitheliochorial placentas are not invasive; in contrast, they might be considered to represent the epitome of the noninvasive placenta. However, in the horse epitheliochorial placenta there are regions, called endometrial cups, where binuclueate cells do invade and erode uterine epithelium. Accordingly, in this section we consider the horse endometrial cups, primarily to elucidate the functional significance of this invasion of uterine tissue and what light it might cast on the significance of the deep invasion of uterine tissue by the human placenta.

In the fifth week post conception, horse trophoblasts form what is termed a chorionic girdle many cells thick. By 35 days post conception (dpc), the apical cells of the girdle become binucleate and begin to express equine chorionic gonadotropin (eCG). The girdle cells invade the uterine epithelium, which is later removed by phagocytosis. The binucleate cells do not divide but rather become large and aggregate into ring formations termed endometrial cups. Interestingly, during the invasive phase, the binucleate cells express MHC (major histocompatability complex) type I antigens that provoke a maternal immune response against paternal antigens. Once these cells have successfully invaded the uterine epithelium and formed the endometrial cups, they cease to express these antigens (Donaldson et al., 1992). The uterine epithelium then grows in from the sides to cover the cup area. The endometrial cups presumably secrete inhibitory factors to keep the maternal lymphocytes from invading and killing them, though these factors have not yet been identified. The maximum secretion of eCG occurs between 50 and 80 dpc; after 80 dpc, eCG production declines, and maternal lymphocytes begin to invade and kill the cup cells. Eventually the uterine epithelium will grow over the scar formed by the necrotic residue from the cups (Wooding and Burton, 2008).

The function of these endometrial cups appears to be solely to secrete eCG into maternal circulation. They do not appear to transport any maternal molecules to the fetus, and in fact, no eCG is found in fetal circulation. Endometrial cups appear to act as a means for the fetus/placenta to signal the mother. The function of eCG is to stimulate progesterone production from the maternal ovaries, thus maintaining the uterus in a receptive state. Failure to establish endometrial cups results in an aborted pregnancy. Thus the ability of the horse placenta to produce endometrial cups that secrete eCG is a requirement for a viable fetus. The invasion of maternal tissue appears to serve a signaling function and not to be directly related to any nutritive function.

Synepitheliochorial Placentas

This placental type is the only one with a phylogenetic basis. It is a modification of the epitheliochorial placenta found in cetartiodactyls but is only found in ruminants. It is best described in cows and sheep (Wooding and Burton, 2008).

Grosser's original classification of ruminant placentas (syndesmochorial) presumed that the uterine epithelium was lost and that trophectoderm was in apposition to maternal connective tissue. Scanning electron microscopy has shown this not to be true. The reality is more interesting and perhaps stranger—fetal binucleate cells fuse with the uterine epithelium to form fetomaternal hybrid syncytial plaques (Wooding, 1984; Wooding and Burton, 2008). Cells form that have both fetal and maternal nuclei! These cells are trinucleate, with two nuclei from the fetus and one from the mother. Somehow, nuclei from different individuals coexist within the same cell membrane and do not stimulate an immune response from potentially antigenic differences in their gene products. These cells secrete fetal hormones and other information molecules into maternal circulation (Wooding, 1992). It is unclear whether both maternal and fetal nuclei are actively expressing gene products. Again, as in the horse placenta, this peculiar adaptation appears to be more involved in signaling than in nutrient or waste transport.

Endotheliochorial Placentas

In endotheliochorial placentation, the uterine epithelium is eroded after implantation, and a fetal syncytium comes into direct contact with maternal endothelium. This is the placental type of almost all carnivores and about 10% of bat species, but it also is the placental type of elephants, aardvarks, sloths, tree shrews, and even a few species of rodents and insectivores (Mossman, 1987). Thus, an endotheliochorial placenta is common among the Laurasiatheria superorder of eutherians but is found in species among all superorders. An endotheliochorial placenta is a successful adaptation for a wide range of species ranging in body size from especially small to especially large, with various habitats and reproductive strategies.

Many endotheliochorial placentas are zonary in form, making a band around the developing embryo. This is true for the carnivores and also for elephants. Most endotheliochorial placentas have a syncytial trophoblast layer; exceptions include elephants, which have a cellular layer.

Endotheliochorial placentas often have hemophagous zones, areas of a placenta where maternal red blood cells are phagocytized and their cell contents absorbed into placental tissue. These zones are thought to be the main source of maternal-to-fetal iron transfer.

Hemochorial Placentas

Humans have a hemochorial placenta. Indeed, all anthropoid primates have this placental type, and so do elephant shrews, most bats, armadillos, rock hyraxes, and tenrecs. Every superorder of mammals contains species with a hemochorial placenta. Even among carnivores, where a zonary endotheliochorial placenta is almost ubiquitous, the hyena develops a hemochorial placenta (Enders et al., 2006). Interestingly, hyenas retain hemophagous zones even though they develop a hemochorial placenta, perhaps an example of mosaic evolution (box 3.1), or perhaps simply retention of an adaptive trait. Perhaps most intriguing, several bat species begin gestation with an endotheliochorial placenta but end it with a hemochorial placenta.

In hemochorial placentas, all three layers of cells surrounding maternal blood vessels are degraded, resulting in maternal blood being in direct contact with the outer placental tissue (usually syncytiotrophoblast). Among hemochorial placentas, distinctions exist as to how many trophoblast layers separate maternal blood from fetal blood, however. There are hemomonochorial (anthropoid primates), hemodichorial (rabbits), and hemotrichorial (rats and mice) placentas, with one, two, and three trophoblast layers, respectively. The functional significance of the different numbers of trophoblast layers separating maternal blood from fetal blood is not well understood. These layers can be cellular or syncytial. For example, in the hemotrichorial placentas of mice and rats, there are two syncytial layers and one cellular layer. In the human

BOX 3.1 MOSAIC EVOLUTION

Animals contain sets of conserved and derived traits. In other words, they possess some ancient traits and others that are relatively new to the lineage. Not all traits evolve at the same rate, or rather, many traits will have been subject to purifying selection, which resists change, and others will have undergone positive or negative selection and thus changed more over time. Generally, animals are a mosaic of ancient, conserved traits and more recent, derived traits. The usual example of mosaic evolution in the human lineage is the contrast between the changes in the pelvis and lower limbs as adaptations to bipedalism that occurred early in our evolution and the morphology of the skull and jaw, which retained the primitive features (robust teeth, small brain case, and so forth) of the common ancestor between us and the chimpanzee. Three million years ago, our ancestors were creatures that walked very much like us but could not think like us.

hemomonochorial placenta, the single layer is composed of syncytiotrophoblast. In some rodents and the rock hyrax, there is no syncytiotrophoblast, but instead their hemomonochorial placenta has only a single cellular layer of trophoblast.

Decidualization

The decidua is the maternal tissue that attaches to the placenta. The original definition of *decidua* was the tissue from uterine epithelium that was shed with the placenta at parturition. The decidua secretes many information molecules and hormones and has receptors for these hormones. As such, it is an important component of the machinery necessary for the maternal-placental-fetal cross-talk that maintains and directs gestational development. Decidualization of uterine epithelium is absent in epitheliochorial placentas and ubiquitous in hemochorial placentas.

In humans and other anthropoid primates, a decidua is formed prior to implantation. This is characteristic of species with a hemochorial placenta. In most anthropoid primates and many bat species, decidualization precedes fertilization. In other mammals with hemochorial placentas, decidual formation is stimulated by the blastocyst. If the uterine epithelial cells do not undergo decidualization, implantation will not be successful. The uterine epithelium needs to be hormonally primed before it can undergo decidualization. Progesterone is an important signaling molecule that cues decidua formation. After ovulation, the uterine epithelial lining is transformed into a secretory lining. If fertilization does not occur, or if the embryo is faulty in ways that result in no or inappropriate signaling, the transformed uterine tissue is shed. In humans, this is menstruation. The regulation of trophoblast invasion into the decidua is discussed in chapter 7.

The Ancestral Placenta

Early conjecture on the ancestral placental form was based on a hierarchical paradigm, influenced by the embryological theories of Ernst von Haeckel (e.g., von Haeckel, 1866). In this hierarchical conception, simple structures are generally considered primitive and more complex structures derived. In addition, those characteristics found in "higher" mammals, such as primates, are considered advanced and derived, especially those features that are found in humans. Thus there was a notion of a progression, from simple to advanced, on which continuum traits fell, with primitive taxa retaining the simpler, ancestral form and more evolutionarily advanced taxa displaying more complex features. Based on this hierarchical thought, the epitheliochorial placenta type was considered the simplest and thus the most primitive. The hemochorial placenta type was considered the most advanced. This was reinforced by the fact that humans and other higher primates had hemochorial placentas, while animals like cows, sheep, and horses had epitheliochorial placentas. There were primates, the prosimians, with epitheliochorial placentas, but these were labeled "lower primates" and considered to be primitive.

This viewpoint has been thoroughly discredited by advances in modern biological knowledge, both in general and for placental evolution. Simple structures can certainly be derived. Structures that humans deem complex may be ancestral. Biological complexity is difficult to determine a priori. Other mammals have had just as many years of evolution as has the human species. All mammals retain some ancestral features and display derived features that are unique to their lineage (mosaic evolution). The placenta is an excellent example of how the hierarchical and progressive viewpoint is misleading and counterproductive. A priori, the human placenta is no more derived than is the horse, cow, or dog placenta. The discredited hierarchical and progressive viewpoint has been replaced by a cladistic viewpoint. The breakthroughs in genome sequencing have given us a better ability to assess phylogenetic relationships on which traits can be mapped, to determine their primitive or derived nature.

An excellent example of how it is difficult to predict what is derived and what might be primitive is the placental development in molosid bats. Initially, a diffuse endotheliochorial placenta forms and supports the fetus through midgestation. At that point, this placenta begins to regress, to be replaced with a discoid hemochorial placenta. Thus, these bat species have the genetic machinery to create two different kinds of placenta. Which is "older"? How does the genetic machinery differ to produce the two different placentas? Do the two placentas rely on different genes for their development, or are the same genes regulated differently to produce the different placental morphologies? At present we cannot answer these questions, but the answers would have important implications regarding which placental type was likely the ancestral one.

Around 130 million years ago, there were animals whose descendants include all of the placental mammal species. That species undoubtedly gave birth to living offspring that were nourished during gestation by a placenta. What did that placenta look like? How would it compare to the diverse placental forms of modern placental mammals? Now that the molecular evidence has given us a phylogeny on which to map placental form and function, this question can be approached.

The Placenta and Phylogeny

Key to reproductive fitness, the placenta is the interface for many significant adaptive interactions. It is also the main organ in which the paternal, maternal, and fetal genomes interact, with all their possible conflicting goals and strategies. Mammalian placentas have undergone extensive adaptive evolution and may well be under the most consistent and strong selective pressure of any organ. In many ways, this fact makes placental structural variation a particularly poor subject with which to investigate phylogenetic relationships. Placental type is generally consistent at the family level but does not seem to assort with known phylogenies at higher taxonomic levels (Enders and Carter, 2004). Many taxa show considerable variation in placental structure, and there would appear to be multiple cases of convergent evolution within mammals. Because we do not as yet understand the genetic underpinning for the differences in placental structures, we cannot assess the likelihood of transformation from one type to another. Is it easier to go from an endotheliochorial placenta to a hemochorial or epitheliochorial placenta, compared with transitioning directly from a hemochorial placenta to an epitheliochorial placenta or vice versa? We don't know, though the fact that primates consist of two groups that are either completely hemochorial or completely epitheliochorial in placental type argues that it is not.

Nonetheless, there have been phylogenetic investigations based on placental type as well as on other morphology and cell structure. One of the goals of these investigations is to reconstruct the most likely placental type of the common placental mammalian ancestor. Earlier investigators, influenced by the discredited notion of progressive evolution and the idea that somehow human biology invariably represents a more evolved state, had postulated that the epitheliochorial-type placenta was the ancestral condition. Endotheliochorial placentas were the next step in this progressive evolution, which culminated in the hemochorial placentas such as found in humans and other so-called higher primates. Of course, hemochorial placentas are also found in armadillos and anteaters, among many other species distantly related to anthropoid primates (Benirschke et al., 2006).

The recent phylogenetic investigations seeking to reconstruct the ancestral placental type have not been able to provide a definitive answer; however, they have agreed that an epitheliochorial placenta is the least likely candidate (Carter and Enders, 2004; Wildman et al., 2006; Mess and Carter, 2007; Elliot and Crespi, 2009). But whether the ancestral placenta was endotheliochorial (Carter and Enders, 2004) or hemochorial (Wildman et al., 2006; Elliot and Crespi, 2009) is a matter of ongoing debate (Wildman, 2011). Investigations at the level of genes might provide better evidence for the constraints imposed by phylogenetic relationships on placental types. There are genes that have largely placental if not placenta-specific activity in many species. Duplications of genes have resulted in gene clusters in which certain of the paralogs have evolved specific placental or gestational function. For example, in rodents it appears that the prolactin gene has been duplicated multiple times, and many of the resulting paralogs have placenta-specific functions. In primates, the growth hormone genes have been duplicated. In many species, retroviral genes have been incorporated into the expressed placental genome. This adds a further layer of complexity that will be discussed in chapter 6.

Phylogeny is not the whole story, of course. Differences in placental structure have arisen within lineages. For example, the prosimian primates (lemurs and lorises but excepting tarsiers) have diffuse, epitheliochorial placentas.

Thus, lemur and loris placentas appear to have more in common with those of dolphins and horses than they do with those of humans, though it is well established that lemurs and lorises are primates, not cetaceans or perissodactyls. And some distantly related species have placental similarities with us; among carnivores, the red panda has a discoid placenta, and the hyena has a hemochorial placenta. Despite these placental similarities, neither of those species is more closely related to us than is any other carnivore. For that matter, the nine-banded armadillo has a zonary, hemochorial placenta with some cotyledonary characteristics. Armadillos are members of the mammalian superorder Xenarthra, which includes sloths and anteaters and is thought to have arisen in South America during the time when that continent was completely separated from North America. Carnivores and cetartiodactyls are members of the superorder Laurasiatheria, which also includes bats; primates are members of Euarchontoglires, which includes tree shrews, rodents, and lagomorphs. A fourth eutherian superorder is the Afrotheria, which includes elephants, aardvarks, and manatees and is believed to have arisen in Africa. Thus, armadillos are equally phylogenetically distant from carnivores, primates, and ruminants but share aspects of placental structure with them all. To further drive home the exceptional spread of placental features across eutherian phylogenetic groupings (clades), within the Xenarthra, a group of mammals that, based on current molecular evidence, are believed to have a shared an ancestor different from those of the other three eutherian superorders, the sloth placenta is classified as cotyledonary and endotheliochorial and the giant anteater placenta as discoid and hemochorial (Bernishcke et al., 2006). Our classification schemes for placentas do not reliably assort species within any accepted phylogeny.

If we look at variation in placental structure among primates and their closest relatives, the best hypothesis is that the diffuse, epitheliochorial placenta found in lemurs and lorises is a derived condition (Martin, 2003; Wildman et al., 2006). The evidence is not definitive, but given that rodents, and more importantly that colugos (formerly known as flying lemurs), the phylogenetically closest living nonprimate species, have discoid, hemochorial placentas, the simplest evolutionary hypothesis is that this placenta type represents the ancestral condition for primates. Selection apparently favored a change from this sort of placenta to a diffuse, epitheliochorial one in lemurs and lorises. There appear to be multiple ways to solve the problem of producing a baby primate, and all placental forms undoubtedly have advantages and disadvantages. In order to understand modern gestational diseases, the disadvantages of our placental type may be quite important to comprehend. There is often an unfortunate, hubristic tendency for scholars to consider modern human morphology as representing some kind of peak biological achievement. Evolution rarely works that way. The genetics that underlie our modern placenta were indeed the most successful within our lineage, but that doesn't mean that its advantages were not accompanied by problems. The fact that selection favored a different path for some, admittedly distant primate relatives, should provide some cause to wonder what challenges provided by a hemochorial placenta were sufficient to nurture such a change and how those challenges might affect placenta-related gestational disease in modern human beings.

4

The Evolution of the Human Placenta

Human beings are descendants of the anthropoid primate lineage. There have been many divergences over the roughly 60–70 million years since the anthropoid lineage came into being. Three major clades of anthropoid primates are generally recognized: the New World monkeys, the Old World monkeys, and the clade that includes all apes and us. Within each of those groups are finer-grained clades that distinguish among the members. For example, among Old World monkeys, there are the cercopithecines and the colobines. Baboons (cercopithicines) and colobus monkeys (colobines) are both Old World monkeys, but they represent different clades of this group. Similarly, depending on the scale of analysis, we as a species can be considered a particular clade within apes or as fitting comfortably within the ape clade as a whole. For example, completely opposable thumbs and shoulder morphology that enables a wide range of movement are features that we share with apes but that separate us (and the other apes) from monkeys. Because of these features we are a firm member of the ape clade. However, adaptations related to bipedalism (e.g. a nonopposable big toe, and aspects of our pelvis and knee joint) separate us from all the other apes. Our large brains, with a significantly expanded cortex, could also be argued to represent a cladistic difference between humans and the other apes.

How does the anthropoid placenta vary across these clades? How unique is the human placenta? Again, the answers to these questions depend on the level of analysis as well as on how the questions are posed. The anthropoid placenta has a number of unusual and even unique characteristics among mammals; the human placenta has its own unique characteristics, as do all primate species. In general, the placenta may be under the most selective pressures of any organ and is thus likely to be variable even between closely related species. An interesting and as yet unanswered question is to what extent the diversity of placental forms is due to different genetics or to differential regulation of common genetics (Wildman, 2011). Gene loci involved in reproductive and immune processes frequently exhibit rapid evolution; considering the importance of the placenta to both reproductive and immune processes, it is perhaps expected that placental gene expression may exhibit rapid evolutionary change among even closely related lineages (Wildman, 2011). Unique genes in several lineages that are largely or solely placentally expressed arose from independent gene duplication events (see chapter 6). We start our investigation looking for both consistencies and differences in placental morphology among primates. We examine especially placental features that are common among the various anthropoid clades.

Primate Placental Morphology

Placentation in primates displays both variability and similarity across taxa. In 1929, J. P. Hill presented a framework identifying four stages of primate placentation (Carter, 1999). In keeping with the times, these stages were presented as an evolutionary scale from primitive to derived, with the human condition considered to be the most derived. However, modern methods and later analyses have demonstrated that some aspects previously considered primitive are most likely secondarily derived (e.g., epitheliochorial placentas in lemuroids and lorisoids). The ancestral placentation type of stem primates is likely hemochorial, or possibly endotheliochorial, but is extremely unlikely to be epitheliochorial (Martin, 2003).

According to Hill, the four stages of primate placentation are lemuroid, tarsioid, pithecoid, and anthropoid. Briefly, Hill argued that the diffuse, nondeciduate, epitheliochorial placentas of lemurs and lorises represented a primitive form of primate placenta. They seemed to represent a simpler form of placentation, with minimal direct contact between maternal and fetal cells. The lemuroid placenta has features in common with that of artiodactyls such as pigs and perissodactyls such as horses. This resemblance to "lower" mammals perhaps further encouraged Hill to consider the lemuroid placenta primitive. Of course the modern conception recognizes that pigs and horses are just as evolved as we are; they just evolved in a different direction, from different ancestral genomes. Pigs and horses are just as divergent from the ancestral placental mammal as are monkeys, apes, and humans.

The tarsioid stage included direct attachment to the uterine wall by the blastocyst and the development of a discoid, deciduate, hemochorial placenta, but it retained some features considered primitive, such as a generally noninvasive implantation and a large, persistent yolk sac. The pithecoid placenta found in

New and Old World monkeys is discoid, deciduate, hemochorial, and invasive; the blastocyst erodes into the uterine epithelium. There are significant differences between New and Old World monkeys in placental features, but Hill considered the similarities in early development sufficient to group them into a single stage, despite the fact that Old World monkeys do share some features of the anthropoid stage. Represented by humans and the living apes, the anthropoid stage is characterized by interstitial implantation, with the blastocyst eroding completely into the uterine epithelium.

Modern phylogentic analysis generally agrees with Hill on the relative relatedness of the taxa characterized by the different placental stages. Tarsiers are more closely related to monkeys, apes, and humans than they are to lemurs and lorises. Humans and apes certainly form a clade within primates. However, the differences in placentation between New and Old World monkeys are consistent with the Old World monkeys being more closely related to apes and humans. The pithecoid stage is probably the least cohesive of Hill's groupings.

Do these stages form a scale from primitive to derived, as Hill suggested? The evidence would suggest not. The closest outgroup to primates (the living creatures that are not primates but are most closely related to primates) are colugos, previously called flying lemurs (order Dermoptera, represented by two genera); colugos have hemochorial placentas. The next closest relatives to primates are the insectivores, which generally have endotheliochorial placentas. Still further away are the rodents and lagomorphs, which have hemochorial placentas. The only other mammalian taxa with epitheliochorial placentas are the perissodactyls (e.g., horses, rhinos) and cetartiodactyls (e.g., pigs, hippos, dolphins), which are members of a completely different superorder (Laurasiatheria) from primates (Euarchontoglires). The most parsimonious hypothesis is that epitheliochorial placentation is a derived condition that has evolved twice in mammals: once in the primate lineage and once in the lineage leading to rhinos, horses, pigs, cows, and whales.

An epitheliochorial placenta has advantages related to the inherent immunological difficulties associated with live birth. There are two epithelial barriers between maternal and fetal blood. This should minimize the chances of fetal cells transferring to the mother, which could cause later immunological issues. It also should minimize the chances of maternal immune attack on the fetus. The immunological advantage of the greater barrier between organisms might translate into a nutritional and metabolic disadvantage, however; it is more difficult for substances (nutrients from the mother, waste products from the fetus, and signaling molecules from each) to pass between the mother and fetus. Of course, these difficulties have been solved. There are a variety of adaptations in species with epitheliochorial placentas that increase the rate of diffusion and transport of molecules between maternal and fetal compartments through various mechanisms, most simply by thinning the barrier. Epitheliochorial placentas are not inherently less efficient; they just employ different strategies and mechanisms to achieve the appropriate transfer rate of materials to and fro. Horses and cows have arrived at independent solutions for placental signaling to the mother: horses by forming invasive endometrial cups and cows by forming trinucleate hybrid cells composed of binucleate trophoblasts fused with uterine epithelial cells. The immunological challenges presented by a hemochorial placenta are not insurmountable; they have been solved in myriad ways by the many species with hemochorial placentas. In the human hemochorial placenta, the maternal immunological system acts as a partner to the blastocyst in remodeling maternal tissue to support implantation.

So, rather than discussing which placental type has advantages of efficiency or signaling or immune protection, we can more fruitfully discuss what the (possibly species-unique) adaptations are that enable each placental type to successfully accomplish its fundamental mission—growing a fetus to become a viable neonate.

A large number of taxa spread across all four superorders of mammals have a hemochorial-type placenta. A hemochorial placenta is found in mammals considered highly primitive (e.g., tenrecs) and highly derived (e.g., humans). This could mean that the hemochorial placenta was the ancestral condition for stem placental mammals (Wildman et al., 2006; Elliot and Crespi, 2009). However, the hemochorial condition is arrived at in a variety of ways (Carter and Enders, 2004); there appear to be many developmental paths to a hemochorial placenta. Examples of species with hemochorial placentas within larger taxonomic groups whose placentas are largely endotheliochorial further suggest that a hemochorial condition is derived. For example, almost all carnivores have endotheliochorial placentas, but hyenas develop a hemochorial condition late in pregnancy. This implies that the widespread existence of the hemochorial placental type probably represents convergent evolution (Carter and Enders, 2004). It also implies that aspects of hemochorial placentation confer sufficient advantages to select for this condition, despite its immunological challenges.

Anthropoid primate placentas can be described as discoid (in some species bidiscoid) and hemochorial. The maternal-fetal interdigitation is villous for Old World anthropoids but trabecular for at least some New World anthropoids (e.g., marmosets; Rutherford and Tardif, 2009). Tarsiers occupy a position of some ambiguity with regard to anthropoid primates. Many scientists consider tarsiers to be part of the anthropoid clade; however, tarsiers are more distantly related to us than is any monkey species (Martin, 2003). Tarsiers have discoid hemochorial placenta with trabecular interdigitation, in common with many New World monkeys.

In the next sections we briefly review what is known about the placentas of selected species from the prosimian, New World monkey, and Old World monkey groups and then provide a more in-depth description of human implantation and placentation.

Galagos

The lesser bushbaby (*Galago senegalensis*) is a small (about 200-gram) nocturnal prosimian primate found in Africa. Typical of the prosimian primates, the lesser bushbaby has a diffuse, epitheliochorial placenta with no decidua (Njogu et al., 2006). The outer layer of the placenta in contact with maternal epithelium is composed of cellular trophoblasts; there is no syncytium or syncytiotrophoblast. The chorionic vesicles opposite the uterine glands are likely involved in maternal-fetal exchange (Njogu et al., 2006). Thus, similar to what occurs in many other species with epitheliochorial placentas, histiotrophic nutrient uptake by the placenta has an important role throughout gestation.

The bushbaby is an example of a small mammal with an epitheliochorial placenta. The bushbaby placenta is significantly different from the anthropoid placenta, and yet it is fully competent to produce small primates that are not greatly different ecologically and metabolically from many small New World primates, such as the common marmoset, which we discuss next.

Marmosets

The common marmoset is a small (about 400-gram) New World monkey native to northeastern Brazil. It is commonly used as a laboratory animal. Marmosets differ from most other anthropoid primates in routinely giving birth to twins; in captivity, triplets and even quadruplets are not uncommon.

The marmoset placenta is discoid and hemomonochorial. The chorion is syncytial, with a discontinuous layer of cellular cytotrophoblasts underneath the syncytiotrophoblast. This is the common anthropoid pattern. Implantation is superficial, as it is in other monkeys.

The placental disks from the twin or triplet fetuses fuse into a single placenta. Thus the fetuses share a blood supply, and they are born hemopoietic chimeras. The differences between placentas from twin and triplet births have been investigated by Rutherford and Tardif. Placentas from triplet pregnancies are qualitatively different and produce more grams of fetal tissue per gram or surface area of placenta, implying some mechanism to increase efficiency (Rutherford and Tardif, 2008, 2009). The same pattern was seen for insulin-like growth factor II (IGF-II) concentration in term placentas, with more grams of fetus produced per gram of IGF-II, implying that at- or near-term triplets are exposed to less IGF-II than are twins (Rutherford et al., 2009), consistent with triplets being smaller than twins.

Embryonic development in the marmoset is unusual for a primate in that it is significantly delayed. Marmosets typically have a fertile postpartum estrus and will often become pregnant within a few weeks of giving birth. The conceptus will implant, and placental growth will commence. Very little embryonic growth or development will occur, however, until about two months post conception. At this time, the placenta becomes endocrinologically highly active, for example, producing copious amounts of corticotropin-releasing hormone (Power et al., 2006). Embryonic development begins, with organogenesis completing soon after. Fetal growth is then normal for an anthropoid primate. This embryonic stasis at the beginning of gestation helps explain why a 400-gram female that produces two (or three) 30-gram infants has a gestation length of 143 days, significantly longer than would be expected for a primate of that small size.

Rhesus Macaques

The placenta is bidiscoid and hemomonochorial in Rhesus macaques, also known as Rhesus monkeys. In general, the definitive structure of the placenta is grossly similar to that in humans. There are some differences, however. Most significantly, implantation is superficial and invasion of the macaque syncytiotrophoblast into the decidua is shallow. The maternal spiral arteries grow through the maternal endometrium to closely appose the minimally invasive trophoblast. In this way, despite the shallow implantation, the placental villi are bathed in a sufficient quantity of maternal blood. As in humans (see below), cytotrophoblasts initially block the maternal spiral arteries, greatly restricting maternal blood flow into the intervillous space during early gestation (Wooding and Burton, 2008).

Humans

The full-term human placenta is compact, discoid, villous, and hemomonochorial. It doesn't start out that way, however—the early human placenta is quite different, in some aspects, from the term placenta. Of course, the challenges the placenta faces are quite different early in gestation compared with near term.

For an embryo to implant successfully requires coordination between blastocyst and maternal uterine physiology. The blastocyst must be activated and the uterine epithelium must be receptive. Cross-talk between blastocyst and uterine epithelial cells is a key component of this process. There are three phases to human embryonic implantation: apposition, adhesion, and invasion (Hannan and Salamonsen, 2007). In all phases, chemical signals travel back and forth between the blastocyst and maternal tissue. During apposition, the blastocyst is close to, but not directly in contact with, uterine epithelium, and soluble information molecules (e.g., cytokines, growth factors) are exchanged. These exchanges are important for the development of both blastocyst and decidua. Adhesion molecules begin to be expressed by both fetal and maternal cells, allowing the next phase, adhesion, to occur. Once the blastocyst has adhered to the uterine epithelium, the exchange of molecular signals can be increased. Implantation in humans (and other great apes) is interstitial—the blastocyst forms a syncytial cap that intrudes through the uterine epithelium, carrying the blastocyst into the endometrium, where it implants in the decidua basalis (invasion phase). Trophoblast cells bud off and invade through the decidua to the spiral arteries. The uterine epithelium that was penetrated by the blastocyst reestablishes itself over the blastocyst, forming the decidua capsularis. Thus, the blastocyst is enclosed within maternal epithelium and isolated from the uterine cavity (box 4.1).

Initially during the first trimester, the placenta surrounds the entire embryonic sac. The exocoelemic cavity separates the amniotic sac from the placenta and contains a secondary yolk sac. Chorionic villi form over the whole of the chorionic sac. The development of the placental bed and the structures for maternal perfusion of the placenta in later pregnancy occur early in the first trimester. Extravillous trophoblast cells infiltrate deeply into the uterine

BOX 4.1 DECIDUA

There are three types of decidua in the human uterus based on their relationship to the placenta and the chorionic villi: the *decidua basalis* (previously termed the *decidua placentalis*), the *decidua capsularis*, and the *decidua parietalis* (also called *decidua vera*). The decidua basalis is the area of actual attachment through which molecules pass between the mother and the fetus. It is in the decidua basalis that the linings of the maternal spiral arteries are eroded, bathing the chorionic villi in maternal blood. The decidua capsularis is the healed uterine epithelium through which the blastocyst burrowed at initial implantation. The chorion is in contact with the decidua capsularis. The decidua parietalis is the uterine epithelium not in contact with the fetal membranes. At about the fourth month of human gestation, the fetus becomes so large that the decidua capsularis is pressed against the decidua parietalis; the two eventually merge, and the uterine cavity disappears.

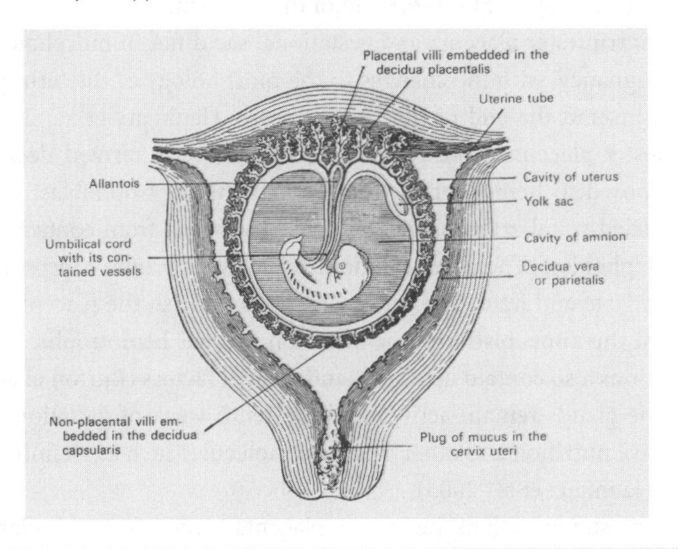

epithelium, effectively eroding all the maternal cellular layers between the placenta and the tips of the maternal spiral arteries in the decidua basalis. The spiral arteries become remodeled to become dilated passive conduits of maternal blood, which flows into the extravillous space, bathing the placental villi in maternal blood. The widening of the terminal segments of the spiral arteries reduces the pressure of the blood inflow, which, among other things, protects

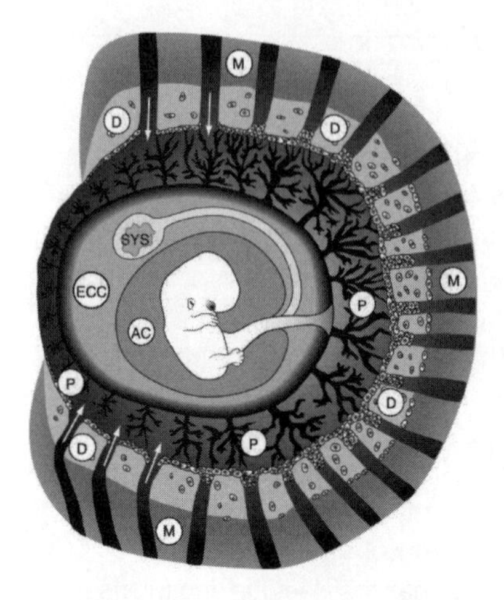

Figure 4.1. Diagram of a gestational sac at the end of the second month, showing the myometrium (M), the decidua (D), the placenta (P), the exocoelomic cavity (ECC), the amniotic cavity (AC), and the secondary yolk sac (SYS). Jauniaux et al., "Placental-related diseases of pregnancy: involvement of oxidative stress and implications in human evolution," *Human Reproduction Update,* 2006, vol. 12, issue 6, by permission of Oxford University Press.

the chorionic villi from shear-force damage. However, this maternal blood flow does not commence until the end of the first trimester.

The first trimester placenta and gestational sac differ in morphology from later in pregnancy, with the changes to the morphology of the term placenta occurring around the end of the first trimester (Jauniaux et al., 2003). The first trimester placental type might be more properly termed deciduochorial as opposed to hemochorial, because extravillous trophoblast cells plug the remodeled spiral arteries, keeping maternal blood from contact with the syncytiotrophoblast. The uterine endometrial glands produce secretions high in carbohydrate and lipid, and these secretions nourish the placenta and the embryo. At this time, histiotrophic nutrition exceeds hemotrophic nutrition. The secretions also contain cytokines and growth factors (Burton et al., 2010). The uterine glands remain active until the tenth week of gestation, providing extensive nutrition and other bioactive molecules such as vitamin B12 and prolactin (Jauniaux et al., 2003).

First trimester growth of the human placenta is remarkably similar among individuals. This implies that the histiotrophic nutrition regime in the first trimester varies little among pregnancies. In contrast, there is considerable variation in placental and fetal growth in the second and third trimesters, when the placenta and fetus are nourished through the maternal bloodstream (hemotrophic nutrition). The variation in growth in the second and third trimesters is often associated with differences in the extent of remodeling of the spiral arteries that occurred in the first trimester. Both intrauterine growth restriction and preeclampsia are associated with shallower implantation (invasion by extravillous trophoblasts) and incomplete spiral artery remodeling.

Growth and development are complementary concepts, but they refer to different processes. Yet, the differences can be subtle, and in many instances the distinction is not only difficult to make but may actually be artificial. In other instances the difference is vital, however, and must be taken into account.

As stated above, for most of the first trimester the invading trophoblast cells effectively plug the ends of the spiral arteries, greatly restricting maternal blood flow into the intervillous space (Burton et al., 2010). This results in low oxygenation of placental and fetal tissue; early fetal development takes place in a hypoxic environment (Jauniaux et al., 2003; Burton et al., 2010). Hypoxia is not conducive to growth per se; it would generally be thought to retard growth. In addition to creating a low oxygen environment, the plugging of the spiral arteries by fetal trophoblast cells restricts the transfer of glucose from maternal to fetal circulation. In all ways, this early, vital period of development would appear to be a time of restricted growth, at least for the fetus. Cell differentiation is the key process that occurs, as well as migration of differentiated cells to specific locations within the embryo. The plan for fetal growth is being produced. Oxygen and glucose are kept at bay, as both can be toxic and teratogenic.

It makes adaptive sense on many levels for the placenta to grow more rapidly than the embryo, at least initially. The placenta is developmentally advanced compared to the embryo. As complex as the placenta is, it is a single organ and not as complex as the fetus will be, with its multiple interlocking organ systems. The placenta must become the first functional organ, and so resources should be initially directed more toward it rather than passed through it to the embryo; this also serves as protection for the embryo. The placenta incorporates glucose, fatty acids, amino acids, and other nutrients into its tissue and uses oxygen in its metabolism, while sparing the early embryo from these necessary but potentially toxic substances. Later in gestation, the fetus will grow at a higher rate than the placenta. The early growth of the placenta allows it to support that greater fetal growth. It would be interesting to know to what extent nutrients are stored in the placenta during early gestation to be transferred to the fetus in late gestation when the maternal supply might otherwise be insufficient.

At about the end of the first trimester, the spiral arteries begin to become unplugged, and maternal blood fills the intervillous space. This shift from histiotrophic nutrition to hemotrophic nutrition presents a challenge to placental tissue. Oxygenation from maternal blood dramatically changes the metabolic circumstances of the placental tissue. Oxidative stress is a significant challenge and results in increased cell apoptosis and cellular damage. It also serves to regulate the placental villous tree structure, causing regression of villi extending into the decidua capsularis.

Initially, chorionic villi form over the whole of the chorionic sac. The onset of maternal arterial circulation commences near the end of the first trimester and begins at the periphery of the implantation site (Wooding and Burton, 2008). This is probably due to the fact that at the periphery of the implantation site, trophoblast invasion is slightest, and thus the spiral arteries are plugged with fewer trophoblast cells. Villi at these peripheral areas show extensive signs of oxidative stress and begin to regress. Eventually, the chorionic and basal plates fuse over the superficial pole of the chorionic sac, forming the chorion laeve. The villi covering the embryonic pole of the chorionic sac continue to proliferate and form the discoid placenta. Eventually two-thirds of the original placenta will regress, leaving a compact discoid placenta in place of the original placenta that completely surrounded the gestational sac.

Oxidative stress, in some ways, appears to serve as a regulatory signal to guide human placental development. It regulates the regression of placental villi away from the implantation site in the decidua basalis. Villous angiogenesis is also influenced by oxygen-dependent growth factors. At low oxygen tension, vascular endothelial growth factor (VEGF) is the dominant growth factor affecting villous angiogenesis. As oxygen tension increases, VEGF is downregulated and placental growth factor (PlGF) is upregulated. Under the influence of VEGF, capillary growth entails the formation of multiple loops, in a branching angiogenesis (Ansari et al., 2005). After PlGF is upregulated, capillary growth occurs by non-branching angiogenesis, leading to the formation of terminal villi.

Fat, the Placenta, and the Fetal Brain

Fat, either as lipid or as represented by adipose tissue, has significant effects on fetal growth and development, maternal and fetal health, and pregnancy outcome. In chapter 7 we will discuss the role of adipose tissue as an endocrine organ, its interactions with the placenta, and the pregnancy complications and birth outcomes associated with maternal obesity. In the present chapter we discuss essential fatty acids, placental transport, and the brain.

Many aspects of our biology set us apart from other primates. However, the human brain represents arguably the most significant set of adaptive changes that distinguish us from all other species. Bipedality is an interesting adaptation, but there are other successful bipedal species. In the past there were species in the genus *Australopithecus* that were, in effect, bipedal apes. They went extinct, but our lineage still exists. Our lineage combined bipedality with a larger brain, and that larger brain enabled other adaptive features of ecology and social behavior to evolve. There is no doubt that the larger brains of our ancestors were a key element in the success of our lineage.

Nothing comes without cost, however. In the biological equivalent of the adage "there is no free lunch," larger brains take more energy and resources to build and to maintain. In 1995 Aiello and Wheeler proposed that the human brain, due to its high metabolic growth and maintenance cost, presented an adaptive challenge to early members of the genus *Homo* (Aiello and Wheeler, 1995). The expensive tissue hypothesis in effect proposed that the Australopithecine dietary ecology was not capable of supporting this extra brain matter; to afford a larger brain, our early ancestors had to shift to a diet of higher-quality foods, which means foods that provide more energy (and possibly other nutrition) per gram. One aspect of a food that generally lowers its quality is being difficult to digest. Leaves generally provide a lower calorie-per-gram return than do fruits and meat, in large part because they are more difficult to digest due to the plant cell wall fraction (fiber). In practice, the dietary shift that *Homo* made early in its existence was to a diet higher in animal matter. This dietary shift, combined eventually with fire for cooking and other methods of food processing that increased food digestibility, enabled our digestive tracts to become less substantial, in effect matching the extra cost of our brains with a decrease in the resources needed by our guts. Our brains got bigger, our guts got smaller, and genus *Homo* prospered.

It takes more than energy, though, to build a brain. Brains are extremely high in fat. This makes them both expensive to grow and also especially good food. Carnivores will often preferentially eat organ tissue, such as liver and brain. But it isn't just any fat that is needed to grow a brain. The brain requires a large proportion of certain essential fatty acids, especially arachidonic acid (AA) and docosahexaenoic acid (DHA). A shift to a more carnivorous diet by our early ancestors likely increased the dietary amounts of AA and DHA (Leonard et al., 2007).

What does all this have to do with the evolution of the human placenta? Well, the larger human brain begins in utero. At birth a human neonate has a brain approximately the size of the brain of an adult chimpanzee. The human brain more than triples in size by adulthood (Ponce de León et al., 2008). Thus most of human brain growth is fueled by mother's milk and then other foods. Still, the human placenta must be capable of supplying the necessary energy and building blocks, especially AA and DHA, to build a brain more than twice the size of a chimpanzee neonate's. In addition, human infants are significantly fatter at birth than many or even most other primates (Kuzawa, 1998). Thus human infants likely are born with significant stores of essential fatty acids, including AA and DHA, which can be applied to brain growth post partum. Did that fatty acid storage require any special adaptive changes to the human placenta? That is a reasonable hypothesis.

Certainly the human placenta efficiently transports AA and DHA from maternal circulation to the fetal compartment. Concentrations of AA and DHA in cord blood are roughly double the concentrations in maternal circulation (Duttaroy, 2009). In a perfusion study using term placentas, increases in the maternal-side concentration of AA, DHA, and α-linolenic acid resulted in twofold-to-fourfold increases in transport (Haggarty et al., 1999). This implies an active transport system, and several fatty acid transport regulatory genes have been identified that are expressed in rat (Knipp et al., 2000) and human (Campbell et al., 1998; Larqué et al., 2006) placentas. However, it is not clear that the human placenta is exceptional in this ability. All brains require AA and DHA. Dolphins with epitheliochorial placentas produce precocious calves with large brains. Dolphins have a diet rich in DHA, as fish is generally a good source of omega-3 fatty acids, including DHA. So perhaps the dolphin placenta does not need to be as efficient. Still, it is best to cast a skeptical eye on assertions of human uniqueness. Dolphins are unequivocal evidence that a hemochorial placenta is not required to produce a large-brained neonate (Martin, 2003).

The Placenta and Human Evolution

Many discussions of the evolution of the human placenta appear to presuppose that our placental type is required, that it is not merely one possible solution among many. One of the questions we ask is, are there aspects of human biology not directly linked to the placenta that require the human placental type (discoid, hemomonochorial, villous, with extensive decidualization, interstitial implantation, and extensive trophoblast invasion), or is our placental type simply one possible solution to the biological challenges of developing a human animal? In one sense, our placental type has uniquely evolved to fit our biology. However, we are descended from species that likely had discoid, hemochorial placentas, probably from the common ancestor of all anthropoid primates. At some point in the lineage leading to apes interstitial implantation evolved, causing more extensive decidualization. It is not clear that the human placental bed differs substantially from that of chimpanzees and gorillas (Carter and Pijnenborg, 2010; Pijnenborg et al., 2011). The evolution of our placenta has been constrained by phylogenetic history as well as by adaptive changes, in response to our particular reproductive strategy. Our placental type is certainly capable of satisfying all the functional needs for producing a successful human neonate. It can feed a large brain and support a long gestation. However, so can an epitheliochorial placenta, as is evidenced by dolphins. An interesting thought experiment is to imagine that whatever adaptive event led to strepsirrhine primates developing an epitheliochorial placenta had happened to catarrhines instead. Would that have made the evolution of the genus *Homo* impossible? Or are there other mechanisms that might have served to solve the challenges inherent in our biology?

Bipedalism

A key feature of our biology that is hypothesized to be an ancestral change from the common ape pattern is bipedal locomotion. Certainly bipedalism preceded the enlarged cortex that characterizes the genus *Homo*. Bipedalism has consequences for pelvic structure, for example, narrowing the pelvic opening through which birth must occur (Rockwell et al., 2003). Bipedalism also changes the relative orientation of the gravid uterus to the bladder, the lower intestines, and several major blood vessels, such as the vena cava. In a quadrapedal animal, gravity pulls the gravid uterus toward the abdominal wall. In a bipedal animal, gravity pulls the gravid uterus down toward the pelvis. During late pregnancy this can compress the bladder, bowel, and more importantly, major blood vessels. This compression of the vena cava and other blood vessels may have provided a selective pressure for mechanisms to improve utero-placental blood flow (Rockwell et al., 2003). The deeply invasive trophoblasts and extensive remodeling of maternal spiral arteries is suggested as, in part, a response to the constraint on blood flow imposed by bipedalism (Rockwell et al., 2003).

The Invasive Placenta

The human placenta invades the maternal uterine wall to a greater extent than is seen in most mammals. Even among anthropoid primates, the invasion of the uterine wall by the human placenta is considered exceptionally aggressive and deep. Yet, an argument can be made that humans do not have the most invasive placenta. That title might be held by the guinea pig, in which trophoblasts can invade beyond the uterine wall and into mesometrial adipose tissue (Roberts, 2010). In addition, what evidence exists suggests that the great apes are similar to humans in their implantation and trophoblast invasion. Certainly implantation in the great apes is interstitial (Carter and Mess, 2010). In both gorillas and chimpanzees, trophoblast invasion of the decidual tissue is deep, extending into the inner myometrium, with extensive remodeling of the spiral arteries, eminently similar to the human condition (Carter and Pijnenborg, 2010; Pijnenborg et al., 2011). It would appear that interstitial implantation and a deeply invading trophoblast with extensive remodeling of maternal spiral arteries is ancestral to at least the African great ape lineage. Thus it definitely predates the enlarged brain of genus *Homo* and predates bipedalism as well, unless bipedalism is an ancestral trait of the African apes that was lost in the gorilla-chimpanzee lineage, a possibility that is not considered likely in current thought on human evolution.

The adaptive function of the extreme invasiveness of the human placenta is uncertain though much speculated on. Several researchers have proposed that it serves to increase nutrient flow to the fetus and that this is an adaptation to support the high metabolic requirements of a large human brain. This hypothesis takes two facts regarding human placental and fetal development, deep invasion by extravillous trophoblasts and a large fetal brain at the end of gestation, and combines them into an adaptive hypothesis. There are several theoretical difficulties with this hypothesis, none insurmountable, but that should serve as a caution against an uncritical acceptance of this idea. From a comparative perspective it is true that the great apes also have large brains, though not to the human extent. An argument could be made that the deep interstitial implantation of great apes served as a pre-adaptation that allowed the further increase in brain size in the human lineage. However, it is not clear that guinea pigs are particularly cephalically endowed, so a deep trophoblastic invasion of maternal tissue may not necessarily relate to brain growth. The most important consideration that urges caution is that there are few if any data that support the concept that increased invasiveness provides for greater transfer of maternal resources or that any such increase is necessary for human fetal brain development. We approach this hypothesis with skeptical but open eyes.

We examine three general hypotheses for the adaptive importance of a hemochorial placenta and a high degree of invasion of the uterine wall by placental tissue in human evolution. The hypotheses can be categorized as: nutritional efficiency, endocrinological signaling, and immunological protection of the infant post birth. They are not mutually exclusive; they may work in concert with each other. A case for all three can be made on evolutionary grounds; they all have potential adaptive significance. An important consideration is what evidence there is to support the existence of proximate mechanisms on which evolution could act. The purpose of this exploration is to investigate the theoretical, and more important, empirical underpinnings of these hypotheses to evaluate their strengths and weaknesses.

Nutritional Efficiency

Physiological data relevant to placental transfer of nutrients, waste products, and signaling molecules is absent for most mammals. Thus comments on the efficiency or lack thereof of different placental types or adaptations are built mostly on theory (or outright speculation) rather than actual data. There are data to suggest that iron transport is more efficient in hemochorial placentas (Seal et al., 1972). Hemochorial placentas generally employ an active transport system involving maternal and fetal transferrins and receptor endocytosis and release mechanisms. In species with epitheliochorial and endotheliochorial placentas, uterine secretions and special hemophagous zones are the main routes of iron transport, but several tenrecs have both a hemochorial placenta and hemophagous zones (Enders and Carter, 2004). Gas exchange, and thus oxygen delivery to the fetus, is unlikely to be significantly different with these various placental types, because it depends more on the thickness of any barrier and not on the number of cellular layers.

Efficiency certainly can be adaptive, but so, under many circumstances, can inefficiency. In part, efficiency and inefficiency become definitions subject to varying standards and assumptions. One measure of placental efficiency is the ratio of the mass of the neonate(s) to the mass (or surface area) of the placenta.

By this measure, the epitheliochorial horse placenta is quite efficient. We say that one placenta is more efficient than another if the relative size between the placenta and the supported fetus is smaller. That presupposes that allocating more resources to the fetus at the expense of the placenta is desirable. But plainly the fetus cannot exist without its placenta, just as a placenta serves no adaptive purpose without an attached fetus. Early in gestation, placental growth exceeds fetal growth—this is adaptive, not inefficient. At the beginning of pregnancy the mother is huge compared to the fetus. The potential rate of transfer of maternal resources far exceeds the fetal ability to metabolize. More is not always better.

This is especially true during early development, when organogenesis occurs. Oxygen is both a necessity and a harmful substance. Oxidative damage is a serious threat to an embryo. Similarly, glucose is both needed and potentially harmful, being teratogenic in early development. Babies of pregnant women with uncontrolled diabetes in the first trimester are at higher risk for a number of birth defects and for fetal death (Reece, 2008). A vital function of the placenta in early gestation is to regulate and, indeed, dramatically reduce the flow of oxygen and glucose from mother to fetus. As mentioned above, during human gestation, trophoblast cells plug the spiral arteries to block maternal blood flow early in gestation.

Still, this phenomenon does not negate the hypothesis that a hemochorial placenta that deeply invades maternal epithelium is important in human gestation. The hypothesis of an essential, greater efficiency in function for the human placenta is only relevant for the end of gestation, when the rapid growth of the brain occurs. Brain tissue certainly does require the delivery of consistent, high levels of oxygen and glucose for proper function and growth. By the end of gestation, the human fetal brain is metabolizing more glucose and oxygen than all other fetal tissues combined. Failure to fully establish the early remodeling of maternal spiral arteries is associated with a number of poor outcomes, including miscarriage, growth restriction in the fetus, and preterm labor (Roberts, 2010).

Do the various placental types vary in their efficiency of nutrient transport? They do vary in their ability to transport certain classes of molecules. However, there are no obvious advantages among placental types regarding total nutrient transfer. A hemochorial placenta perhaps presents the fewest difficulties, due to the direct contact of trophoblasts with maternal blood. Hemochorial placentas may have an advantage regarding iron transfer, though the hemophagous zones found in many non-hemochorial placentas are an adaptation to increase iron transfer and thus an alternative adaptation to this challenge. In several bat species, the initial placenta is endotheliochorial, but by mid-gestation that original placenta begins to regress, to be replaced by a hemochorial placenta. The timing implies that the hemochorial placenta is more efficient and is adaptive for supporting the higher nutrient requirements of the later-term fetus. However, all placental types are capable of producing large, precocious offspring.

To evaluate fully the hypothesis that the human placenta adapted to our larger brain requires going beyond a consideration of hemochorial versus other placental types. All anthropoid primates have hemochorial placentas and all have a large brain relative to most other mammals; however, humans have exceptional brains even among the anthropoids. So in what ways does our placenta differ from other anthropoids' that might affect brain growth? Does interstitial implantation and a deep invasion of decidual tissue offer any advantages for efficiency of nutrient transport or ability to transport larger fluxes of nutrients? Would the macaque and baboon pattern of superficial implantation and shallow invasion of the decidua be incapable of supporting human fetal requirements during late gestation? It is difficult to answer that question. However, gross examination of the intervillous space and maternal blood flow in baboons and macaques does not imply any particular deficit in comparison to the human condition (Wooding and Burton, 2008). It is not clear that the human placental bed inherently provides greater nutrient transport than does that of macaques.

Signaling

If transport of nutrients hasn't been a limiting factor in human placental evolution, then perhaps the answer lies with signaling molecules. Many regulatory molecules are produced in the placenta and secreted into maternal circulation to influence maternal metabolism and physiology (e.g., growth hormone, placental lactogens, corticotropin-releasing hormone). The various placental types have different capacities to signal the maternal system, influenced by the interhemal distance, the number of maternal cell layers between maternal blood and placental trophoblast, and evolved mechanisms for active transport. Bioactive molecules also go from mother to fetus, and again transport is related to placental type. Molecules produced by fetal organs can influence placental function and even be passed into the maternal circulation, sometimes after being metabolized by the placenta. For example, dehydroepiandrosterone sulfate (DHEAS) is produced by the fetal adrenal gland in humans and other anthropoid primates, converted into estrogen by the placenta, and then passed to the maternal circulation (Kallen, 2004; Rainey et al., 2004a, b).

In general, the assumption is that placental signaling (in all directions: fetal-to-maternal, maternal-to-fetal, and placental-to-maternal) is the greatest with hemochorial placentas, followed by endotheliochorial, with epitheliochorial last. As with all generalities, there are exceptions for particular signaling pathways, but the basic concept appears reasonably supported. The fewer maternal cell layers between maternal blood and the outer tissue of the placenta, the better signals are able to pass back and forth. A hemochorial placenta is not required for placental-maternal signaling that regulates maternal metabolism; however, a hemochorial placenta enables that kind of signaling to occur without specialized adaptations.

A number of anthropoid primate–specific signaling molecules are produced by the placentas of anthropoids. The hemochorial placental type allows easy access to the maternal circulation for placental signaling molecules. Anthropoid primates certainly have evolved a set of signaling molecules produced by the placenta and secreted into the maternal blood—placental lactogens, from duplications of the growth hormone gene, and chorionic gonadotropin, from a duplication of the lutenizing hormone gene. Anthropoid primates are also the only species known that produce placental corticotropin-releasing hormone, which has significant effects on maternal metabolism. These molecules are absent from the lemurs and lorises, with their epitheliochorial placentas. The evidence supports the hypothesis that a hemochorial placenta is capable of supporting greater placental signaling to the mother. Yet, as the evidence from horses and ruminants has shown, there are solutions to the challenge that enable an epitheliochrial placenta to adequately signal the maternal physiology. An investigation of prosimian primate placental signaling might identify adaptations with a similar function.

The final consideration is whether the interstitial implantation and deep invasion by extravillous trophoblasts in human beings produces any advantages in maternal-placental signaling. Is there any evidence that placental signaling in humans is any greater or more potent than signaling in macaques? We are not aware of any. The macaque syncytiotrophoblast is bathed in maternal blood just as is the human syncytiotrophoblast.

Immunoglobulin Transfer

The protective role of maternal immunity for human infants was first documented in 1846. During a measles outbreak on the Faroe Islands, it was noted that if a pregnant woman survived the disease and delivered her baby, the baby appeared to be immune to measles. This phenomenon was also seen with smallpox vaccination, where vaccinating pregnant women protected their babies from smallpox after birth. In modern medicine there are a number of vaccines recommended for use during pregnancy, with the expectation that they will not only protect the woman but also provide some level of inherited immunity for her child when born (Englund, 2007).

This is not a human-specific phenomenon. In several other species, mothers are known to transfer immunoglobulins, specifically IgG, to their fetus(es) during gestation, thus transferring information about the mother's immune history to her offspring in utero and conferring some acquired immunity to disease before birth. However, present evidence indicates that placental IgG transfer only occurs in species with hemochorial placentas. Specifically, there is no evidence that immunoglobulins can transfer with either epitheliochorial or endotheliochorial placentas. Thus, the one aspect of a hemochorial placenta that appears not to be duplicated by other placental types is the ability to transfer immunoglobulins (at least IgG types) from the mother to the fetus. Other aspects—nutrient transfer, signaling, and metabolic regulation—appear to be achieved regardless of placental type, although hemochorial placentas may have some advantages. Epitheliochorial and endotheliochorial placentas have an immunological advantage over hemochorial placentas in that the maternal bloodstream is buffered to a greater extent from paternal antigens, and the fetus is buffered to a greater extent from maternal antigens. Hemochorial placentas offer both immunological advantages and disadvantages.

Maternal immunoglobulin transfer across the placenta to the fetus requires crossing two barriers in anthropoids: the syncytiotrophoblast and then the fetal endothelium. Transfer is accomplished by an active transport process using receptors for IgG, called Fc receptors because they have an affinity to the Fc fragment of the IgG molecule. These receptors are found in both the syncytiotrophoblast and in fetal endothelial cells (Mishima et al., 2007).

Not all anthropoid placentas appear equal in their ability to transfer IgG to the fetus (Coe et al., 1994). Newborn squirrel monkeys (New World anthropoid primates) had IgG levels equal to only about 35% of maternal values. In contrast, in Rhesus monkeys and chimpanzees, IgG levels in newborn infants were roughly equal to maternal levels. The highest levels of IgG were found in human babies, where newborn infant values were actually significantly higher (139%) than maternal values, supporting the finding of an active transport system for IgG in the human placenta (Kohler and Farr, 1966). In a prosimian primate, the galago *Otolemur garnetii*, IgG levels in newborn infants were barely detectable, at 3.5% of maternal levels (Coe et al., 1994). The galago has an epitheliochorial placenta and thus would not be expected to employ immunoglobulin transfer before birth.

Beyond the evidence for placental transfer of IgG, there is evidence that maternal transfer of immunoglobulins has been increased in the human lineage. Human milk contains high concentrations of IgA, far higher than concentrations measured in macaque milk (Milligan, 2005). So it would appear that both in utero and post partum, human infants receive higher transfers of maternal immunoglobulins (IgG and IgA) than do infants of our nonhuman primate relatives. Why should an increased capacity for immunoglobulin transfer in human beings have evolved?

The relatively recent (in an evolutionary sense) advent of agriculture and the domestication of animals during human history provides a possible answer. Agriculture allowed an increase in population density and long-term occupation of living sites. Settlements became permanent, with multiple generations living in the same place. Sanitation issues arose; there would have been more possible vectors for disease (for example, other humans), and the concentration of parasites and pathogens in the immediate environment would have increased, unless sanitation methods were employed. Novel pathogens would have been brought into the environment along with domesticated animals. Many of our most significant diseases originate from outside our species—witness bird and swine flu of recent fame. In many cases, infants will be the most susceptible to and suffer the greatest morbidity and mortality from disease and parasites. The evidence of our placenta and our milk suggests that pathogens affecting infancy were significant selective forces in our evolution. A hemochorial placenta pre-adapted the human lineage to be able to upregulate IgG transfer. If anthropoids had evolved an epitheliochorial placenta, as did lemurs and lorises, then it is reasonable to assume that passing prebirth immunity from mother to child would not have been possible. Whether or not a hemochorial placenta is necessary to support the fetal pattern of human brain growth, it did allow a protective immunological adaptation to the eventual consequences of our having a large brain.

Is the greater maternal transfer of IgG related to the deep, interstitial invasion of trophoblasts found in human pregnancy? It does not seem likely. Transfer of IgG across the placenta is an active transport process. Differences in transfer rate probably reflect adaptive changes to the transport system. For example, a potential hypothesis is that the human placenta has a greater density of Fc receptors than do other anthropoid placentas. Chimpanzees did not differ from Rhesus macaques in the fetal concentration of IgG, implying equivalent maternal transfer; and yet, the chimpanzee placental bed is similar to the human and different from the macaque. Implantation in chimpanzees is interstitial, and trophoblasts deeply invade the decidua and fully remodel the spiral arteries. There is no evident difference from the human condition.

Comparative Advantages of Placental Types

All placental types exhibited by extant eutherian species are, by definition, successful. Producing a viable neonate presents a broad range of biological challenges to which there are many adaptive solutions. The solution at which any particular species has arrived depends on the ancestral condition, the particular selective pressures the species has undergone, and a random aspect, the available mutational changes selection acted on. Thus any species' placenta will have characteristics that are conserved within its phylogenetic relatives and, at the same time, is likely to have idiosyncratic differences due to both selection and random accumulation of mutations. In addition, placentas of many eutherian species appear to have accumulated functional genes from retroviral infections (see chapter 6). This adds to the complexity—differences between placenta form and function of phylogenetically distant species may be influenced by genes coming into a lineage from outside.

From examining the evidence, what can we conclude about the hypothesis that the human placenta adapted, by necessity, to the fundamental changes leading to our evolution: bipedalism and a large brain? A hemochorial placenta is the ancestral condition in anthropoid primates, so that does not represent a change. Interstitial implantation and the deep invasion and remodeling of the spiral arteries appear to be ancestral conditions for great apes (though the latter is not certain for orangutans), so again that does not represent a unique human adaptation. These conditions may relate to a large brain, as great apes have larger brains for their body size than other anthropoids (except for the New World monkeys in the genus *Cebus*, for which there are no data on placentation). However, the increased size of the human brain does not appear to require any significant change from the great ape condition. Neither, it would appear, does bipedalism. It is possible that further study of great ape placental bed morphology and physiology will detect differences that could relate to either bipedal locomotion or the enhanced encephalization of our lineage, but more likely the invasive placenta phenotype is either at best a preadaptation that allowed bipedalism and the human brain to expand or possibly not related to locomotion or brain size at all.

5

Sex and the Placenta

The diversity of placental structure, along with certain findings from molecular biology research, attest to the intense selective pressures this organ has been subjected to within the many mammalian lineages. It is an organ that has experienced considerable adaptive evolution. The placenta also appears to have affected basic genetic mechanisms within mammals that serve to regulate and drive mammalian biology. For example, among animals the phenomenon of gene imprinting appears to be mainly a mammalian genetic mechanism to guide offspring development.

The placenta is obviously intimately associated with sexual reproduction. Although the placenta is derived from embryonic tissue, it can be considered an adaptation for female reproduction. However, the paternal genes are also represented in the placenta. The placenta has been suggested as a major site for maternal-paternal-fetal conflict, in the evolutionary sense. Of course, it is also a site where cooperation among these distinct genomes toward complementary fitness goals occurs.

In this chapter, we examine sex determination and its effect on genetic regulation mechanisms in the fertilized egg and eventually in the placenta. We end the chapter with a discussion that starts to consider the phenomenon of gene imprinting and how existing mechanisms of DNA repair, silencing, and regulation of expression have been adapted to enable gene expression determined by gene origin. This powerful genetic regulatory mechanism appears to exist predominantly, if not solely, in mammals and flowering plants, two highly divergent taxa sharing a reproductive mechanism that allows intimate contact between maternal and offspring genomes.

Sex Determination

How do individuals become male or female? Sex differentiation begins during embryonic development. There are key time periods when the developmental pathways can be perturbed from their genetically programmed trajectory by hormones that pass through the placenta. This can affect both sex phenotype and sexuality. In other words, early life events can affect both the initial gonadal and other organ structures and the later developmental patterns and environmental-developmental interactions that result in sexual behavior. Sex differentiation is not the same as sex determination. Sex differentiation is the suite of developmental processes that result in the development of the gonadal, brain, and other organ structures that create phenotypic sex and later sexuality. Sex determination is the process by which sex differentiation is guided onto either the male or female pathways (Sarre et al., 2004).

Sex determination in vertebrates falls into two basic categories: genetic or environmental (e.g., temperature-dependent). Among genetic mechanisms, the existence of sex chromosomes is fairly widespread; however, there are sex-determining genes that are not on a sex chromosome per se. In some taxa (e.g., fish, birds) there are even sex reversals in individuals, where a female will become male or vice versa.

Although it is not yet known what the ancestral condition for sex determination in vertebrates was, educated guesses abound. Some researchers have emphasized that the two categories of sex determination (genetic and environmental) really represent opposite poles of a continuum (Sarre et al., 2004). In some taxa (for example, reptiles) both mechanisms are important (e.g., Shine et al., 2002). There are even species in which both mechanisms operate, as described below for the Australian three-lined skink.

Temperature-Dependent Sex Determination

It has long been known that incubation temperature has a dramatic effect on the sex ratio of hatchlings of all extant crocodilians, many turtles, and some lizards. There is also a family of birds (megapodes) in which the sex ratio of hatchlings is known to be temperature dependent. These birds do not sit on their eggs; rather, they bury their eggs in mounds that they construct out of leaves, dirt, and sticks. For example, the Australian brush turkey builds large compost mound nests in which the females bury their eggs. The eggs are warmed by the heat of decomposition. The males tend the nest, checking the temperature by thrusting their beaks into the mounds and adding or removing material. At temperatures around 34°C (the mean for mounds measured in the wild) equal numbers of male and female chicks hatch. At cooler temperatures, more males than females hatch, while at warmer temperatures there are more female chicks (Göth and Booth, 2004).

Multiple possible mechanisms can work to achieve variation in sex ratio of offspring due to incubation temperature. For example, in addition to temperature-dependent sex determination, there could be differential mortality between females and males at different temperatures, or even sex reversal. In the Australian brush turkey, sex is determined genetically, as in all birds; differential mortality accounts for the various sex ratios of hatchlings at different temperatures (Göth and Booth, 2004). So the proportions of males and females successfully hatched from a nest, but not the sex of any individual, are determined by temperature. However, in an Australian lizard (*Bassiana duperreyi*, the Australian three-lined skink) sex is determined by both genetic and temperature-dependent mechanisms. Three-lined skinks have heteromorphic sex chromosomes (i.e., sex chromosomes of different sizes). Females are the homogametic (XX) sex and males are heterogametic (XY). However, when eggs were incubated at low temperatures (consistent with nest temperatures during cool summers at higher altitude in the wild) males accounted for more than 70% of hatchlings (Shine et al., 2002). As mean nest temperature increased, the sex ratio approached 50:50. Differential mortality could not account for the differences in sex ratio between temperatures; it appears that at cool temperatures, some XX individuals became phenotypically male (Shine et al., 2002). Of course, they can only produce sperm with X chromosomes and therefore only genetically female (XX) offspring.

The mechanisms by which incubation temperature directs the developmental process toward one sex or another are not understood. There probably are multiple mechanisms that operate differently in various species. Although the cues and mechanisms by which development is directed toward the male or female path are highly variable among species, the actual developmental pathways appear highly conserved (Crews, 2003; Sarre et al., 2004). Many of the genes that initiate and support the developmental cascade that results in either female or male sex are found in all vertebrates. Thus, it would appear that the pathways to male or female are ancient and highly conserved, but the triggering mechanisms that start an individual down one path or the other are highly variable and have undergone substantial evolution.

Live Birth and Sex Determination

As we argued in chapter 2, one advantage of live birth is greater control over the environment in which the embryos will develop. In mammals, this includes the temperature regime. Mammals are homeothermic animals; in other words, mammals maintain (reasonably) constant internal body temperatures. Reptiles are heterothermic; their body temperatures vary with the external environment. Even so, it is quite possible that reptiles can regulate their body temperatures to achieve a more stable temperature than would be found in a nest. At first glance it would seem that reptiles that produce live-born young would have genetic-based sex determination. For example, all snakes are ZZ male and ZW female, and a large number of snake species produce live-born young. However, viviparous lizard species exist that have temperature-dependent sex determination. Intriguingly, in one such species, there is good evidence that females regulate their basking behavior to produce sex ratios of offspring in response to their perceptions of the social environment, producing more males if adult males appear to be rare and more females if adult males appear to be common (Robert et al., 2003). In this case, behavioral temperature regulation has become an adaptive mechanism to regulate the sex ratio of offspring.

Sex Chromosomes

In mammals, phenotypic sex is largely determined by genetic factors. The phenotypic sex of an individual (though not necessarily the sexuality of the individual) is determined at conception. Most people are familiar with the placental mammalian sex chromosome system: XX for females and XY for males. Sex determination in marsupial mammals is the same, females being XX and males XY, but the monotremes are different. The platypus has ten sex chromosomes, arranged in five pairs. They act like an X and a Y chromosome in that they assort in two sets of five (X_{1-5} and Y_{1-5}). During meiosis the X and Y chromosomes form chains. Thus a gamete (egg or sperm) has either five X chromosomes (egg or sperm) or five Y chromosomes (sperm only). A female platypus is $X_1X_1X_2X_2X_3X_3X_4X_4X_5X_5$ and a male is $X_1Y_1X_2Y_2X_3Y_3X_4Y_4X_5Y_5$. The other extant monotremes, the echidna species, are similar, with a slight twist—their five X chromosomes are matched by only four Y chromosomes (Rens et al., 2007).

Convention divides sex determination by sex chromosomes into two categories: XX female and XY male in systems where males are the heterogametic sex and ZZ male and ZW female if females are the heterogametic sex. This convention is used even though the Z and W chromosomes may or may not have anything in common with the X and Y. For example, the bird Z and W chromosomes share greater homology with human autosomes (e.g., Z shares large regions of homology with human chromosomes 5 and 9 and the human

X chromosome is more homologous to chicken chromosome 4) than they do with either the X or Y chromosomes (Rens et al., 2007; Veyrunes et al., 2008). The Z and W chromosomes of snakes are not homologous to the Z and W chromosomes of birds (Ezaz et al., 2006; Matsubara et al., 2006; Kawai et al., 2007). To add some additional complexity to the evolution of sex determination in mammals, the sex chromosomes of the platypus share greater homology with the sex-determining chromosomes of birds (Z and W) than they do with the X and Y chromosomes of other mammals (Veyrunes et al., 2008); this occurs despite the fact that male platypuses are the heterogametic sex, as in all other mammals, while in birds the females are heterogametic.

Mammals all fall into the X and Y category, even though monotremes are a bit unusual in having multiples of each. Birds and snakes are Z and W sex determined. Amphibians are a mix; some are XX female and XY male, while others are ZZ male and ZW female. Most impressively, the Japanese wrinkled frog (*Rana rugosa*) has populations where females are XX and males XY and other populations where males are ZZ and females are ZW (Ogata et al., 2008), though in this case, the chromosomes are undoubtedly much related to each other (Nakamura, 2009). In this frog, the Y and Z chromosomes are virtually identical, as are the X and W chromosomes; the different terminology is created by convention and is not an indication that the sex chromosomes have completely changed. Exactly how these different combinations produce male and female frogs is not yet fully understood. Somehow the sex-determining genes have moved between chromosomes in the different populations of this species.

The evolution of the different sex chromosome systems is not fully understood. Because of the low homology between the mammalian and avian sex chromosomes, many researchers have suggested that the XY and ZW systems evolved independently. The mammalian XY chromosome system is thought to have derived from a pair of autosomes over the last three hundred million years (Ross et al., 2005). The current accepted phylogenetic tree has the mammalian lineage diverging from the reptile and bird lineages between 300 and 350 million years ago. Reptiles and birds diverged from each other about 250 million years ago. Sex determination in reptiles is variable. There are taxa with ZW sex chromosomes similar to birds (e.g., snakes), but there are also species in which males are the heterogametic sex (e.g., most lizards). In all crocodilians, many turtles, and tuatara (lizard-like reptiles that arose about 250 million years ago; only 2 species exist today), sex determination is driven by the incubation temperature and is not dependent on sex chromosomes at all. Thus the sauropod lineage that represents most amniotes displays all the mechanisms for sex determination. The high degree of homology between platypus and chicken sex chromosomes suggests the possibility that the ZW system, with males being homogametic and females heterogametic, may have been the ancestral condition.

Amphibians are the tetrapod outgroup to the sauropods (e.g., reptiles and birds) and synapsids (mammals). In other words, amphibians are equally distant phylogenetically from all of the amniotes but are more closely related to amniotes than is any other taxa. To evaluate the origins of mammalian and avian sex chromosomes, comparisons with amphibians would be helpful. One such study, using a salamander, has shown that both the mammalian X and avian Z chromosomes share substantial homology with one of the salamander autosomes (Smith and Voss, 2007). Therefore, it would appear that the sex chromosome systems of mammals and birds probably did derive from a common ancestral chromosomal sex-determining system, originally derived from an ancient tetrapod autosome. Whether the ancestral system was XX/XY or ZW/ZZ cannot be determined.

The Human X Chromosome

The human X chromosome is composed of an ancient, highly conserved region (XCR) that is common to most if not all placental mammal X chromosomes. It is highly conserved with regard to the marsupial X chromosome (Graves, 2008). However, it also contains a relatively recently added section that derives from an autosome (Graves, 2010). The X and Y chromosomes pair during meiosis only in short regions at the tips of the arms of the X chromosome, termed the pseudoautosomal regions.

Approximately 1,100 genes have been annotated for the human X chromosome, with another 700 pseudogenes. This gene density is among the lowest for annotated human chromosomes. About 10% of the expressed genes on the human X chromosome belong to the cancer-testis (CT) antigen group. These are genes predominantly and often solely expressed in testes but also in various cancer cells. Thus, the human X chromosome is highly enriched in what would appear to be genes that might confer male advantage or be essential for male development and function (Ross et al., 2005). The X chromosome also has a significant number of genes related to brain function (Graves et al., 2002).

The X chromosome has many long repetitive sequences, especially long interspersed nuclear elements of type 1 (abbreviated LINE1), comprising about one third of the X chromosome (Ross et al., 2005). These LINE1 elements are important for X inactivation; they appear to form a framework or scaffold that guides where the noncoding RNA from the *Xist* gene (see below) attaches and thus blocks transcription of the covered DNA (Carrel et al., 2006).

Consequences of Sex Chromosomes

In humans, the X chromosome is about three times the size of the Y chromosome, as measured by total DNA. In terms of numbers of functional genes it is even larger, with more than 1,000 expressed regions (genes) versus 178 transcribed DNA sections on the Y, of which only 45 represent unique proteins (Graves, 2006). The human Y chromosome contains large segments of repetitive sequences, including simple repeats that have no coding function and also appear to have no phenotypic consequences (Graves, 2006). In many birds, the female W chromosome is significantly smaller than the Z chromosome, with about one-tenth the DNA (Namekawa and Lee, 2009). Some researchers have proposed that evolution favors degeneration of the heterogametic sex-determining chromosome. Some even predict that eventually it will be lost, doomed to degenerate completely. Predictions of when the human Y chromosome will be lost range from 14 million to as few as 125,000 years from now (Graves, 2006). Plainly, the genes critical to male determination and function would have to survive somehow, so either the theory is flawed (e.g., there may be a minimal size after which the Y chromosome will cease to have a net loss of genes) or the key, required genes will be copied to an autosome, and perhaps the system will begin anew with a different pair of sex chromosomes.

Heterogametic sex chromosome degeneration is not universal. In many amphibians, the different sex chromosomes do not vary much in size. In ratites (ostriches, emus, and rheas), the Z and W chromosomes are effectively the same size. On the other hand, there are species in which the Y chromosome apparently has degenerated completely and vanished. Grasshopper males are XO (O represents no chromosome pairing with the X), and several rodent species as well (mole voles and Ryukyu spiny rats) appear to have lost the Y chromosome (Graves, 2006, 2008).

In therian mammals and most birds, the grossly unequal amounts of genetic material between the homozygotic and heterozygotic sex chromosomes have several implications. The first to consider relates to meiosis. In species with heteromorphic sex chromosomes there is a theoretical problem during meiosis; the problem is merely theoretical because, of course, all extant species with heteromorphic sex chromosomes have, by definition, solved it. During meiosis the chromosomes form pairs (meiotic synapsis). In male mammals and female birds, the small size of the heterogametic sex chromosome results in incomplete, or possibly even nonexistent, pairing during meiosis. An ancient DNA preservation mechanism usually results in the silencing of unpaired DNA. The sex chromosomes of the heterogametic sex in both mammals and birds are silenced during meiosis, such that in sperm of mammals (and ova in birds) the sex chromosomes are at least temporarily incapable of transcription. This inactivation is of short duration in birds but lasts longer in mammals.

X Chromosome Inactivation

It has been known for over 50 years that random inactivation of one of the X chromosomes in each cell occurs in female eutherian mammals. This is the underlying genetics of the fur colors of tortoiseshell or calico cats. These cats are overwhelmingly female (rarely, a phenotypically male cat is actually XXY and thus can express the tortoiseshell phenotype). A gene that determines cat coat color is located on the X chromosome and has a black allele and an orange allele. In male cats with only one X chromosome, coat color will be, at base, either black or orange. Unquestionably, other genes influence coat color as well, so that the large diversity of coat colors and patterns seen in cats can be expressed. But only in cats with two X chromosomes in each cell nucleus can there be a mixture of black and orange colors.

In 1949, Murray Barr and his graduate student Ewart George Bertram reported on the existence of a different chromosomal structure seen only in the nuclei of female mammalian cells (Barr and Bertram, 1949). This modified chromosome became known as a Barr body. In 1959, Susumo Ohno, an important American geneticist and evolutionary biologist, published a paper showing that Barr bodies were in fact condensed and heterochromatic X chromosomes (Ohno et al., 1959). Mary Francis Lyon proposed that this inactivation of the X chromosome in female mammals explained color coat variation in mice (Lyon, 1961). Independently, Ernest Beutler showed that African American women heterozygous for glucose-6-phosphate dehydrogenase deficiency (an X-linked trait) had two types of red blood cells in their circulation: one red blood cell type expressed the gene for glucose-6-phosphate dehydrogenase, and the other did not. Both of these findings extended Ohno's observation of X inactivation in female cells to a concept of female mosaic X chromosome expression. Thus, female eutherian mammals have only one X chromosome in each cell capable of gene expression, but the active X chromosome can be different in different cells. Patches of cells derived from a common ancestor cell early in development will all have the same version of the X chromosome inactivated. Thus a female eutherian mammal is a mosaic of cells expressing either the maternal or paternal X chromosome.

Why do female mammals silence one of their X chromosomes? What adaptive function comes from not expressing genes? The basic theory behind why one of the X chromosomes should be inactivated in a cell was that otherwise, females would express twice the amount of proteins, RNAs, and other X chromosome gene products than would males. Since a single X chromosome must be able to express sufficient gene products for male survival, and excess can often be as deadly as deficiency, the theory is that having two expressed X chromosomes in a cell would probably be lethal. Or at the very least, female metabolism (broadly defined as all the chemical processes of life) would of necessity be in a notably different balance of X-linked and autosomal signals and processes than would the metabolism of males. By inactivating one of the X chromosomes, female metabolism faces the same relative chromosomal signals as does male metabolism. Certainly many genes exist where excess production will have profound and often detrimental effects. There are probably relatively few genes where the appropriate range of basal expression is large enough to encompass a 100% increase in production.

Doubtless, X inactivation is not the only method to achieve this balance. For any particular gene product, there could be other means to regulate expression, but chromosome inactivation is the only global method. How well does this theory hold up, given a broad examination of genetic sex determination in animals? It certainly isn't obvious that the mammalian system is required. Even within mammals, there is diversity in the specifics of the X inactivation system; when birds, snakes, butterflies, and fruit flies are considered, it becomes clear that the placental mammal adaptation to the problem of gene expression in sex chromosomes is not the only possible one (Hyunh and Lee, 2005). Birds do not show global inactivation of the Z chromosome in males (Ellegren, 2002). Indeed, for some genes there appears to be no dosage compensation, with males producing twice the amount of transcribed RNA as do females (Itoh et al., 2007). However, for other Z-linked genes, expression levels in male birds are lowered, such that gene expression from two Z chromosomes is approximately equal to expression from the single Z of females; for many other genes, expression is intermediate, with males having gene expression higher than females' but lower than twice that of females'. A mirror-image strategy is to upregulate sex chromosome expression in the heterogametic sex. For example, fruit flies show an approximately 100% increase in X chromosome expression in males compared to females (Luchessi et al., 2005; Heard and Carrel, 2009). The monotremes, with their complex sex chromosome sets, are similar to birds, as some genes show compensation, either full or partial, while others show no compensation (Deakin et al., 2009).

Marsupials versus Eutherians

We have been emphasizing that in placental mammals, a female is a mosaic of active X chromosomes, with some cells expressing the paternal X and some the maternal X. Marsupials differ from placental mammals. In all marsupial mammals so far studied, it is always the paternal X that is inactivated (Deakin et al., 2009; Okamoto and Heard, 2009). This is an interesting difference that needs to be carefully considered and explored to understand the evolution of placental mammals from an ancestor shared with the marsupials.

The marsupial X chromosome has considerable homology with the placental mammal X chromosome; about two-thirds of the gene content is the same (Deakin et al., 2009). Since marsupial and eutherian mammals diverged about 150 million years ago, this is a remarkable degree of chromosome gene content conservation. Given the high degree of homology between marsupial and eutherian X chromosomes, it is reasonable to hypothesize that these two lineages face similar gene dosage compensation challenges. It is also reasonable, though less so, to predict that marsupials and eutherians have met these challenges using common mechanisms. Inarguably, X chromosome inactivation occurs in marsupials. But the details differ, and not just in that paternal X as opposed to random X chromosome inactivation is the rule (Deakin et al., 2009; Okamoto and Heard, 2009).

Although there is remarkable conservation of gene content between marsupial and eutherian X chromosomes, gene order is not as conserved. The arrangement of genes differs between marsupial and eutherian X chromosomes; this difference appears to derive from a greater amount of gene rearrangement in the marsupial lineage. The eutherian X chromosome appears to have undergone few gene rearrangements; gene order is remarkably conserved, given the many tens of millions of years separating placental mammal lineages. In contrast, despite the overall conservation of gene content, the marsupial X chromosome has undergone substantial gene rearrangement. Ohno (1967) suggested that gene order in the eutherian X chromosome has been preserved as a consequence of the global gene compensation mechanism, X inactivation. Furthermore, on mouse and human X chromosomes, concentrations of long interspersed nuclear elements (LINE-1s) correlate with the degree of inactivation (Deakin et al., 2009). These concentrations of LINE-1s have been proposed as acting to enhance the inactivation process (Carrel et al., 2006). The opossum X chromosome does not have LINE-1 concentrations, implying that LINE-1s are not important to marsupial gene inactivation mechanisms (Mikkelsen et al., 2007). Thus, marsupial and eutherian mechanisms to inactivate one X chromosome are qualitatively different, even if these mechanisms serve to address some of the same adaptive challenges.

The *Xist* Gene

Eutherian mammals have an X inactivation center (XIC) on the X chromosome (Brown et al., 1991). A gene (*Xist*) located near the center of the mammalian X chromosome codes for a large nontranslated RNA molecule. This RNA molecule coats the X chromosome and begins the inactivation cascade. Genes are inactivated over time, starting from the *Xist* gene and radiating outward in both directions (Chaumeil et al., 2006). Thus, in early development there is a range of gene inactivation on the X chromosome that changes over time as more genes become inactivated. The LINE-1s that are spread throughout the X chromosome appear to enhance this process of gene silencing (Carrel et al., 2006; Deakin et al., 2009).

Marsupials do not have a *Xist* gene. The genes that comprise the XIC in placental mammals are not grouped together in marsupials (Hore et al., 2007). However X inactivation is accomplished in marsupials, the details of its regulation are different from those in placental mammals.

When does X inactivation occur? There is some uncertainty; in therian mammals, inactivation and reactivation appear to occur over several time points. In placental and marsupial mammals, both sex chromosomes will be inactivated during meiosis for sperm production in the male. Therefore sperm carry either an inactivated Y chromosome or an inactivated X chromosome. This would fit with the global paternal X inactivation seen in marsupials. Certainly by the morula stage in mammals, the paternal X is inactivated. In the eutherian embryo, during epiblast formation the paternal X chromosome is reactivated, and then random X inactivation occurs. The trigger and mechanism for the reactivation and then random inactivation is not known. In rodent models, this process does not occur in all embryonic cells, only in the inner cell mass that is destined to form the fetus. Of note, in the outer cells that will form the trophoblast, the placental tissue that forms the interface with maternal tissue, it is the paternal X chromosome that remains inactivated, at least in murid rodents (Takagi and Sasaki 1975; Wake et al., 1976; West et al., 1977) and possibly in cows (Dindot et al., 2004).

Marsupial embryos have no inner cell mass. There are certainly cells that will form the basic, short-lived placenta (trophoblast) and cells that will form the fetus (pluriblast), but they do not segregate into an outer and inner layer during development as they do in placental mammals. All cells retain their paternal X chromosome inactivation, in common with the outer cell mass that will form the trophoblast of the placenta in rodents. It is reasonable to hypothesize that this pattern was the ancestral condition (Deakin et al., 2009).

Could there be any particular advantage to paternal X chromosome inactivation in the placenta? Some researchers have speculated that paternal X chromosome inactivation in trophoblasts serves to reduce the number of expressed paternal antigens, decreasing the immunological complications of the intimate contact between the trophoblasts and the maternal uterine epithelium. It is possible that the ancestral condition was global paternal X chromosome inactivation, as in marsupials. When the developmental step of an inner cell mass destined to become the fetus arose in placental mammals, the selective pressure for global paternal X inactivation could have been relaxed. Random inactivation of the X chromosome in the fetus allows greater variation in gene expression. Is this, on balance, a selective advantage? Inarguably it is successful in at least some placental mammals.

Is this pattern the rule in placental mammals? The data are not clear. In rodents so far studied it does appear that, in general, one X chromosome is activated and one inactivated in every cell of a female fetus and her placenta. In the fetus the inactivation is random; in trophoblasts it is at least highly skewed toward paternal X chromosome inactivation, if not completely paternal X inactivation. However, in humans the pattern differs; the human placenta is a mosaic of random X chromosome inactivation (de Mello et al., 2010), similar to the condition of the fetus.

Have Some X Chromosome Genes Escaped?

Biology is often messy. It is rarely certain and complete. Variability and even a modicum of inherent unpredictability appear to be hallmarks of biological systems. Inactivation of the X chromosome is a good example. A simple story becomes more and more complicated as it is further investigated and better understood. In humans it appears that some X chromosome genes are doubly expressed in female tissue. About 15% of X-linked genes appear to escape inactivation, and another 10% are incompletely inactivated (Brown and Greally, 2003). Genes that have been (relatively) recently translocated to the X chromosome from an autosome are more likely to escape inactivation. This suggests that the inactivation process proceeds over generations. Newly translocated DNA probably lacks the markers that enable the methylation or histone modifications that will silence transcription. Those markers will likely accumulate over generations, if silencing transcription is beneficial.

Epigenetics

Epigenetics is an old concept that precedes our modern understanding of genetic mechanisms. The term and original concept were introduced by C. H. Waddington (1942), combining epigenesis from embryology with genetics, to describe how genotype interacting with environment gives rise to phenotype. Holliday (1990, 449) defines epigenetics as "the study of the mechanisms of temporal and spatial control of gene activity during the development of complex organisms." A more recent conception of epigenetics is heritable changes in gene expression without underlying changes in DNA sequence (Jones and Takai, 2001). The modern definition of epigenetics is narrower and refers primarily to mechanisms that silence genes (e.g., DNA methylation) without changing the underlying DNA structure (Crews and McLachlan, 2006).

The concept is perhaps most easily understood in the regulation of gene expression to produce different tissues or specific cell types. All cells in our body have the same inherent genetic potential. Heart cells do not differ from kidney, brain, or gonad cells in terms of DNA sequence; instead, gene activation varies between cell types. Some genes that are expressed in the heart are silenced in the gonads and vice versa. One mechanism to achieve this is DNA methylation, which generally acts to silence a gene; genes with methylated promoter regions are inactivated (Holliday, 2006).

Epigenetic change is stable with respect to cell division and can even be passed from generation to generation. For example, an epigenetic mutation in the plant *Linaria vulgaris* changes the symmetry of the flowers from bilateral to radial. This mutant was first described by Linnaeus more than 250 years ago. The mutation results from methylation of the *Lcyc* gene, silencing it (Cubas et al., 1999). However, the original genetic potential has not been lost; flowers will occasionally revert to wild type through demethylation of *Lcyc*.

Epigenetic regulation of the genome relies on somehow marking sections of DNA for modification after the sex cells (egg and sperm) have combined. Histone modifications and methylation of DNA appear to be the two most commonly employed mechanisms.

Gene Imprinting

In the case of autosomes, most genes are expressed by both chromosomes. The phenomenon of X inactivation is an understandable deviation from the normal pattern. Sex chromosomes and autosomes are inherently different. Paternal X inactivation does, however, make it quite clear that there are mechanisms by which chromosomes, or more specifically segments of DNA, can be marked as coming from one parent or the other. Markers passed through meiosis and into the gamete can inform the eventual development of the embryo by favoring the expression of DNA from one parent or the other. All DNA is not equal.

There are genes on autosomes in which only one copy will be expressed (monoallelic expression), and the copy that is expressed is not random but rather depends on which parent provided the gene. In other words, either the gene from the paternal or the maternal chromosome will be expressed. This is termed "gene imprinting," in a quite loose analogy to the phenomenon of behavioral imprinting, where newly hatched chicks will behaviorally imprint upon the first animate object they see after hatching. The terminology for epigenetic imprinting can be a little confusing. If a gene is paternally imprinted, it means that the paternal copy of the gene is epigenetically modified such that it is *not* expressed. Thus, with a paternally imprinted gene only expression of the gene inherited from the mother occurs, and for a maternally imprinted gene, expression is only from the paternally inherited copy. In other words, the term *imprinted* is associated with silencing rather than with expression.

Imprinted genes are an important aspect of mammalian development. Both marsupials and placental mammals express many imprinted genes, especially in the placenta. Monotremes, however, do not appear to have imprinted genes, though it is possible that they just have not yet been discovered (Bartolemei, 2009). Certainly the genes commonly found to be imprinted in placental mammals are not imprinted in the platypus. So gene imprinting is not a definitive mammalian trait, but it would appear to be a therian (metatherian and eutherian) mammal trait.

Other amniotes do not seem to use gene imprinting. So far, it has not been found in reptiles or birds. However, interestingly, it is a common finding in a vastly different organism: flowering plants (angiosperms). In angiosperms, a successful seed results from two fertilization events (a pollen grain contains two sperm nuclei), one for the seed and one for the endosperm that surrounds the seed and provides it with nourishment. Imprinted genes have been identified, in almost all cases in the endosperm. The endosperm is usually triploid, containing a diploid maternal genome and a haploid paternal genome. Imprinted genes thus have either biallelic maternal expression or expression by the single paternal gene copy. The endosperm serves to feed the developing embryo and is analogous to the placenta, providing a connection between the maternal genome and the developing offspring. Indeed, some of the placental products for sale are actually derived from plant "placenta," that is, endosperm.

Imprinted genes have significant consequences for mammalian biology. For example, parthenogenesis is not a viable reproductive strategy in mammals. Perhaps one in a thousand vertebrate species are parthenogenic (Mittwoch, 1978), and true parthogenesis in vertebrates appears limited to squamate reptiles (Sinclair et al., 2010). There are a number of all-female lizard species. It has long been known that parthogenesis in mammals results in nonviable embryos (Mittwoch, 1978). Gene imprinting provides an explanation for why parthenogenesis is lethal in mammals. Appropriate development of the placenta and the fetus is dependent on certain genes being expressed from only the maternal or paternal genomes. Of note, genes that are paternally expressed tend to be associated with placental and fetal growth, while many maternally expressed genes are associated with brain development (Davies et al., 2005, 2006).

Maternal-Fetal Conflict

Fundamentally, this is a book about mammalian female reproduction. Female mammals bear the brunt of the physiological, immunological, and metabolic consequences of reproduction. Although the placenta is derived from embryonic tissue, it can be considered an adaptation for female reproduction. The mass and energy required to grow a placenta and fetus ultimately come from maternal tissues. The placenta forms the interface between the maternal and fetal tissues; the placenta sends signals to, and receives signals from, the mother. The uterus must make adjustments in coordination with the extraembryonic tissues of the placenta, or implantation will not be maintained or even occur. Thus, the structure and function of the placenta must be compatible with maternal physiology. The maternal genome has a significant effect on placental growth and development. The placenta affects and is affected by maternal physiology. Information is continually exchanged between the mother, the placenta, and the fetus, mainly in the form of bioactive molecules. The placenta is highly influenced by the molecular cross-talk with maternal tissue. And of course, half of the placental genome derives from the mother. The placenta is, in part, a female reproductive adaptation, even though it is genetically other.

However, males are a necessary part of mammalian species; at the least, sperm are required. Paternal DNA comprises half of the nuclear DNA of the placenta and fetus (mitochondrial DNA are largely maternally inherited). The placenta is perhaps the most significant interface for the interactions between paternally derived genes and maternal resource allocation and final reproductive effort directed toward her conceptus. And, obviously, one half of a female's offspring will, on average, be male. Male genes must function within the maternal-placental-fetal unit. Male mammal reproductive fitness is dependent on a functional fetal-placental-maternal interaction.

In all species, there is potential for conflict and cooperation between maternal, paternal, and offspring genomes. The terms *conflict* and *cooperation* are human analogies; they rather neatly, though not perfectly, summarize the concept that the fitness outcomes of various reproductive adaptations are not identical for all parties. In addition to the fitness consequences for the mother, father, and individual offspring in question, consequences regarding sibling fitness are also involved, which clearly feed back onto maternal and paternal fitness as well (especially as all the siblings may not have the same father). From a genomic perspective, the fitness consequences of the biology of any individual can be quite complicated.

In general, the fitness imperatives of parents and offspring will largely coincide. Producing offspring that survive to maturity is a major component of reproductive fitness for both of the parents, and obviously it is a necessary condition for the offspring. As the old saying goes, however, the devil is in the details. The optimal outcome for mothers, fathers, and offspring probably will not be identical. Variation that improves the overall reproductive fitness of all the participants, parents and offspring, will have high selective advantage. In terms of evolutionary theory, traits that benefit both parent and offspring are perhaps not terribly interesting, though in terms of understanding the adaptive reproductive biology of a species, they can, and likely will be, vital. Traits with this characteristic will spread rapidly through the population and likely become the norm, if not completely fixed traits. In contrast, traits that benefit one or more, but not all of the actors in the reproduction play, will have a more uncertain fate. Doubtless, there will be variation in traits that benefits one of the players to the detriment of the others. Traits of this kind are of immense interest to theoretical evolutionary biologists. Conflict often generates more interest than cooperation.

In mammals, it is generally the case that females endure greater physiological loads and constraints due to reproduction. Once a female mammal becomes pregnant, her future reproductive options have been significantly constrained. This is not necessarily true for a male who has impregnated a female—in some species that male's reproductive efforts can continue largely unabated. Not in all cases, assuredly, but nonetheless an evolutionarily significant asymmetry exists.

It has long been understood that males and females will often have different fitness goals and thus different and sometimes conflicting reproductive strategies. The same kinds of conflict and tension can occur between parents and offspring. "Parent-Offspring Conflict" was the title of an influential work by Robert Trivers (1974) that encouraged the systematic examination of the evolutionary consequences of divergent fitness strategies and goals between reproductive adults and their progeny. Placental mammals add an intriguing complexity to the fitness interactions among fathers, mothers, and offspring. The placenta not only provides a connection between the mother and fetus through which to pass nutrients and waste products; it also provides a connection by which maternal and paternal genes can interact. Maternal gene products influence fetal and placental growth, development, and metabolism, just as products of the fetal genome can affect maternal biology. The placenta affects maternal physiology and vice versa. Information is continually exchanged between the mother, the placenta, and the fetus, mainly in the form of bioactive molecules. The placenta is genetically identical to the fetus, but plainly the expression of that genome is fundamentally different. However, although the placenta is derived from fetal tissue, it is highly influenced by the molecular cross-talk with maternal tissue. The placenta thus becomes a principal organ by which conflicting fitness goals of all the players (paternal, maternal, and fetal) can be expressed (Haig, 1993).

Parental investment was defined by Trivers (1972, 139) as "any investment by the parent in an individual offspring that increases the offspring's chances of survival (and hence reproductive success) at the cost of the parent's ability to invest in other offspring." Importantly, the cost to investment in other offspring could be for current or future offspring. Thus, not only can the new baby take time and attention away from older siblings, it can also affect the parent's ability to invest in any future siblings. Parental investment as defined by Trivers is a powerful and subtle concept but also one that can be intrinsically hard to measure, especially the ability to invest in future offspring. When used in practice, shortcuts are often taken in an understandable attempt to produce predictions that are testable by experiment or observation. Most costs are assumed to represent investment, often with careful theoretical justification, but sometimes without a serious consideration of why a particular cost might actually impose a Triversonian investment. The use of the word *investment* is also somewhat misleading, because people have in mind the more common conception of investment in terms of finances, which differs in important ways from the parental investment model. Investment is thought of as resources put toward some endeavor, with the expectation of future gain. That notion certainly fits with the idea of reproductive investment (to use a broader term); resources are devoted to behaviors, anatomy, and metabolism with the expectation that these "investments" will, on average, increase reproductive fitness. However, the Triversonian conception is that investment means resources put into offspring that can lower future reproductive fitness, a subtly different idea. Surely in many cases, the two formulations of investment are the same. Indeed, anything that would be considered investment under the Trivers model would also be considered investment under the more general, financial system model. But the reverse is not true. A mother can invest resources in an offspring with no effect on her ability to offer resources to future offspring. Not all costs are Triversonian investments.

In this book we restrict our discussion to maternal investment relevant to the placenta, which to a large extent restricts the kinds of investments to metabolic investments. In other words, nutrients, costly signaling molecules, and other biochemical maternal resources are the investments in question, and only maternal investments exist. Fathers have a stake in the matter but actually contribute nothing but genetics. Mothers and their offspring are actively engaged in signaling and responding to signals in order to regulate the flow of maternal resources to the fetus through the placenta. Fathers are engaged secondhand, by the effects of their genetic contribution to the placenta on signaling and regulation. Gene imprinting is hypothesized to be one outcome of the back and forth of genetic conflict and cooperation between maternal and paternal fitness objectives (Moore and Haig, 1991; Reik and Walter, 2001; Wilkins and Haig, 2003; Haig, 2006).

Gene imprinting and X chromosome inactivation may be linked both conceptually and mechanistically. Just as imprinting appears to be a mammalian adaptation, so does X inactivation. In nonmammals, other mechanisms can achieve dosage compensation due to differing numbers of sex chromosomes. Global sex chromosome inactivation, like gene imprinting, appears to be a therian mammal adaptation. Haig (2006) has proposed that both genomic imprinting and X chromosome inactivation arose in mammals due to genes having asymmetric effects in matrilines and patrilines. The fundamental mammalian adaptation of lactation, followed later by placentation, has increased reproductive costs for females and created a large asymmetry between female and male fitness strategies. According to Haig's model (Haig, 2006) this asymmetry would favor mechanisms to silence genes by parental origin. The placenta would be an organ of particular importance in the expression of this maternal-paternal antagonism. The fact that in cows, mice, and rats paternal X inactivation appears to be the rule in the placenta suggests that paternal X inactivation, as seen in all tissues so far examined in marsupials, is likely the ancestral condition. Random X inactivation is a derived condition.

Since gene imprinting and X inactivation are present in both marsupial and placental mammals, but neither appears to be active in monotremes, the question of which came first cannot be answered. If they have common origins, at least in some of the mechanisms for identifying parent of origin, these two processes have evolved independently as well. The evolution of true viviparity, with a placenta forming an intimate connection between fetus and mother, may have been a fundamental factor in the evolution of these two, relatively unique genetic regulatory mechanisms in mammals.

6

Genes, Genetic Regulation, and the Placenta

In 1983 Barbara McClintock (figure 6.1) was awarded the Nobel Prize in Physiology or Medicine for her work on the cytogenetics of maize. In particular, her work on transposons, the "jumping genes," helped give her this well-deserved recognition. The discovery of the phenomenon of transposition, the ability of segments of DNA to insert themselves into novel locations on chromosomes, in essence to jump about the genome, fundamentally changed the perception of the structure of the genetic material underlying inheritance. Chromosomes were not static, slowly changing structures, subject only to rare mutational events and the occasional recombination; rather, they contained dynamic, fluctuating elements that could dramatically increase the rate and frequency of change.

By 1948 McClintock had already established herself as a recognized expert in maize cytogenetics. Her early research was on visualizing chromosomes. She developed techniques that enabled the morphology of the ten maize chromosomes to be seen. Harkening back to our comments in chapter 1, scientific advances were again being enabled by technical advances, but also by clever experimental design. Still, the focus was on the structure of the genetic material, what could be seen. This is not a criticism of McClintock's work but rather a statement regarding the nature of our species, in which visual information has been of great evolutionary importance and static, structural phenomena appear easier for our minds to comprehend than dynamic, regulatory ones. An advantage of science, however, is that it builds on itself. Through her techniques, McClintock was able to link groups of traits that were inherited together to the specific chromosome where they were located (still structural). However, working with a graduate student, Harriet Creighton, McClintock demonstrated the phenomenon of chromosomal crossover and showed that it was the process by which genetic recombination occurred (Creighton and McClintock, 1931). Although this did not directly dispute the prevailing ideology of genes as "pearls on a string," it did demonstrate that the organization of genetic material was mutable and that inherent processes, beyond rare

Figure 6.1. Barbara McClintock, winner of the Nobel Prize in Physiology or Medicine in 1983 for her work on the cytogenetics of maize. Courtesy of the Barbara McClintock Papers, American Philosophical Society.

mutational events, affected genetic change. This was an early start on a more dynamic understanding of genetic material.

In the late 1940s, McClintock made a pivotal discovery that was to influence her research efforts and her theoretical ideas about genetic regulation for the rest of her life. She discovered that two genes in the maize genome (which she termed Activator, or *Ac,* and Dissociator, or *Ds*) moved about on chromosome 9 (McClintock, 1950). Through a series of careful, technically competent, and clever experiments, she was able to show that although *Ac* appeared always able to move about the genome, *Ds* was only able to move in the presence of *Ac*. In cells without *Ac, Ds* was fixed in place. These jumping genes most definitely did not obey the "pearls on a string" model of genes. Genes were not necessarily fixed in their location on the chromosomes. More importantly to her, however, jumping genes could affect the expression of other genes. When *Ds* moved into the location of a known gene, that gene was silenced, and when *Ds* moved from the location of a gene, that gene was again expressed. Here, apparently, was a mechanism to regulate gene expression.

McClintock understood that regulatory mechanisms had to exist in order to translate the structure of inheritance into the process of development. After all, the genetic information in heart cells is identical to the genetic information in cells from other tissues, such as kidney, skin, and brain. How does the same DNA produce such different cells? There must be mechanisms to regulate the expression of the DNA in order to produce the variety of cellular form and function that exists within a single organism. McClintock correctly perceived the importance of figuring out these genetic regulatory mechanisms. Although much of her work was on the structure of the chromosomes, the overarching goal of her research was focused on understanding genetic regulation. She built a theory of genetic regulation around the action of the jumping genes. She called them "controlling units" initially and then renamed these jumping genes "controlling elements." She did this deliberately to set them apart from normal genes. What was important in her mind was not the fact that these segments of DNA could move around the genome but that their movement affected the expression of other genes.

This theory was not well received. Indeed, there are popular descriptions of its reception as "stony silence" (Keller, 1983) or at best indifference. A perception exists that McClintock was not appreciated during most of her life and that it was only later in life, after other (male) researchers had confirmed her findings in other species, that she was given the credit she was due. These popular accounts often ascribe the poor reception of her ideas to her sex and possibly her personality, which was not particularly social and outgoing. She was something of a recluse and was quite happy to be alone with her work. She never sought the spotlight; instead she was uncomfortable in it.

These perceptions that McClintock was slighted by the scientific community of her time are not fully in accord with the recognition that same community accorded her during her career, however (Comfort, 2001). In 1944, at the age of 42, she was elected to the National Academy of Sciences, one of the highest honors for a scientist. The following year she was elected president of the Genetics Society of America. She was a well-respected member of the genetics community. She was invited to present at prestigious meetings, such as the 1951 symposium where she presented her theory of controlling elements.

McClintock's work on the structure of chromosomes, recombination, chromosome breakage, and jumping genes and her final publication in 1981, "The Chromosomal Constitution of Races of Maize," were all of the highest quality and contributed to the advancement of knowledge. However, her theory of controlling elements was, to put it bluntly, wrong. Transposons are not a mechanism for genetic regulation. Instead, they are a mechanism for increased genetic diversity and novelty to arise within a lineage. They are extremely important to understanding evolutionary processes and genetic change, but they are not fundamental to cell differentiation or genetic regulation. In this one aspect of her work McClintock was searching for the right answers but wasn't looking in the right place.

McClintock wasn't wrong in concept, however; there are definitely molecules coded by DNA sequences that could be considered controlling elements, regulating the expression of the genome. For example, micro RNAs (miRNAs) are short sequences of RNA, coded by DNA but not translated into proteins. Rather, these miRNAs bind to complementary strands of messenger RNA (mRNA) transcribed from other coding DNA sequences and thus either silence or enhance the translation of the mRNA. Other mechanisms to regulate DNA transcription include histone modifications of DNA and DNA methylation, both of which rely on the actions of enzymes such as DNA methyltransferases, coded for by DNA. So there are genes, both protein-coding and RNA-coding, that have global effects on gene regulation. All these mechanisms and more are processes that regulate genome expression and are important to the differential expression of the genome, which can result in cellular differentiation. In typical fashion, scientific progress has followed technological progress. McClintock didn't have the tools to investigate the gene-regulating mechanisms that have now been shown to operate. If she were alive today, we expect that she would be very comfortable with, and indeed excited by, the advances in knowledge concerning genetic regulation. She would probably be contributing to progress in understanding gene regulation and cellular differentiation using the more powerful tools of modern biology.

The Placenta and the Genome

The patterns of gene expression in the placenta are unique. They have to be, since the placenta is genetically identical to the fetus but obviously quite different in development and function. Of course, that is true of all the organs of the body; they are genetically identical, but the expression of the genome differs markedly. The placenta is, in effect, the first organ that develops, an extraembryonic organ. The fetus becomes a baby and starts an independent existence. The placenta dies and is generally thrown away, or possibly eaten.

Modern biology is equipped with tools far beyond those of the early placental investigators. These tools have allowed scientists to broaden their investigations of the placenta beyond structure. Anatomy and structure still matter, both at the tissue and cell levels, but immunology and endocrinology have appropriately risen as key areas of research into placental function. Regulation has become the key concept, as opposed to structure, or perhaps in addition to structure. In other words, we can now move beyond the functional anatomy and descriptions of cell types to explore further the means by which the placenta regulates maternal and fetal physiology and metabolism through information molecules (cytokines, hormones, and so on) and even through genetic regulation of the fetus. Scientists are beginning to explore gene expression and regulation in the placenta in order to understand how it develops and how it functions.

In this chapter we discuss aspects of genetic regulation and gene expression in relation to placental function and evolution. We consider the functional significance of repeated DNA sequences, both long and short, and their association with transposons and retrotransposons. Jumping genes appear to have been of great significance in shaping the mammalian genome. Retrotransposons and retrovirses share functional organization that implies that they are derived from each other. Certainly some retrotransposons are the remnants of past retroviral infections; whether the reverse has happened, where a retrotransposon has managed to become a retrovirus, is a fascinating and scary thought. The placenta has been shown to have important associations with repeated sequences and retrotransposons. Doubtless there are ancient retroviral genes that perform important functions in the human placenta. Repeated sequences and retrotransposons are associated with DNA methylation. The DNA of placental cells is typically hypomethylated compared to that of fetal cells. The placenta would appear to be a hot spot for expression of normally silenced DNA sequences. Finally, we consider the timing of gene expression in the placenta and the role of gene duplication in placental signaling function. First, however, we examine some basic questions regarding genome size and organismal complexity.

Genes, Life, and Complexity

We are slowly beginning to understand how the genome functions to create complex organisms composed of highly differentiated organ systems. Intriguingly, the complexity and diversity of life does not arise simply from a corresponding diversity of genes. The number of expressed genes does not vary dramatically among different species, at least not to the same extent as humans perceive the variability of complexity among living things. *Caenorhabditis elegans* is a millimeter-long nematode (roundworm) that has been used extensively as a laboratory model organism. It was the first multicellular organism to have its genome sequenced. Its genome consists of about 100 million base pairs and contains a little over 19,000 protein-coding genes and about the same number of noncoding RNA genes. The genome of the fruit fly (*Drosphila melanogaster*) has also been sequenced; it has 165 million base pairs, an estimated 14,000 protein-coding genes, and at least as many noncoding RNA genes. Thus the fruit fly does not seem to exhibit greater genetic complexity compared with the nematode, though humans tend to perceive an insect as a more complex organism than a one-millimeter-long roundworm. In comparison, the mouse and human genomes contain 2.5 and 2.9 billion base pairs, respectively, and each has about 30,000 protein-coding genes. The number of RNA genes is more difficult to determine but is probably roughly equivalent. Thus mammals do appear to have more DNA (about 20 times more) and more genes (though only about 1.5–2 times more) than do roundworms and insects. The more interesting aspect of this finding is that the amount of DNA increases faster than the number of protein-coding genes. Scientists are beginning to look within the noncoding genome for answers to biological complexity.

The original, somewhat naive view of the relation between genome size and organismal complexity was that more complex organisms should have larger genomes. A human was thought to be more complex than a frog or a fly and certainly a nematode. Therefore, it should take more genetic instructions to produce a human. But in gross genome-size terms, this relationship did not always hold. The largest genome known so far is that of a slow-growing plant (*Paris japonica*) found on the Japanese island of Honshu. At 150 billion base pairs, this plant has 50 times the amount of genetic material as human beings (Pellicer et al., 2010). Bullfrogs as well have larger genomes than do humans. Birds have relatively small genomes compared to other vertebrates (Gregory, 2002; Hughes and Piontkivska, 2005). Unquestionably, most flying animals have relatively small genomes compared to nonflying members of their lineage (Gregory, 2002). Bats have smaller-than-average genomes among mammals (Smith and Gregory, 2009), and the nonflying birds have larger-than-average genomes among birds (Hughes and Friedman, 2008). Smaller genomes translate into smaller cells, which not only translate into less overall weight but also into a higher surface-area-to-volume ratio for cells, which enhances gas exchange and thus allows higher metabolic output. Both of these outcomes are advantageous to flying animals. So there does appear to be at least one selective advantage to a trimmed-down genome. But why do some apparently simple organisms have such large genomes?

One proposed solution was that much of the DNA in a genome is actually nonfunctional; in other words, it is "selfish" or junk DNA that has managed to be replicated in various lineages without actually performing any function. It does neither good nor harm to the organism but merely exists and replicates. In that case, the important number to be correlated with organismal complexity would be the number of expressed genes, DNA that produces a functional product.

Undoubtedly, there is junk DNA within most genomes. In the human genome, only about 1.5% of the DNA is known to actually encode a protein. It does not seem reasonable that 98.5% of human DNA is nonfunctional. When the mouse genome was sequenced, large regions of noncoding DNA were found to be conserved between mice and humans. That discovery implies significant adaptive function for that "junk" DNA, in order for it to be conserved over 75 million years of independent evolution. Although only a tiny percentage of a genome might encode proteins, as much as 80% of DNA is transcribed into RNA. We now know that noncoding RNA genes are not always merely so-called junk DNA; many noncoding sequences have functions. They don't accomplish their purpose by producing a protein, but they are functional nonetheless. Importantly, many noncoding sequences regulate the expression of coding genes. For example, the miRNA strands that bind to DNA work by either increasing or decreasing DNA expression. Many other small RNAs act on messenger RNA and even on the chromatin. The *Xist* gene for X chromosome inactivation, assisted by long interspersed nuclear elements (LINE-1s) that were previously assigned junk status, produces an RNA transcript that massively silences gene expression. Incontrovertibly, the mammal genome contains an exceptional number of repeated sequences, likely derived from transposons and retrotransposons that have probably been co-opted to perform regulatory functions.

Biology's regulatory nature extends naturally into the genome. Not only does the genome contain the information that regulates the making of the organism, but the genome itself is regulated by mechanisms both inherent to DNA and by cellular factors that interact with DNA. The range of known transcription factors acting to regulate gene expression has greatly increased. The cellular environment is full of molecules that bind to DNA regions and either enhance or suppress gene expression. Ligand-receptor complexes, peptides, and miRNA are among the known DNA transcription regulators. Methylation of DNA and histone modifications are additional mechanisms that affect DNA transcription and greatly increase the complexity of possible gene expression patterns. Robert Waterland, working in Randy Jirtle's laboratory with mice, demonstrated that nutritional factors acting through the mother during gestation can directly affect gene expression in her offspring. If during pregnancy, yellow agouti mouse dams are given high levels of the nutrients choline, betaine, folic acid, and vitamin B12, all of which act as methyl donors, the resulting offspring show increased methylation of DNA, specifically of a transposable element upstream of the agouti gene, that changes the gene expression and thus the phenotype of the pups (Waterland and Jirtle, 2003). Other researchers have shown that low folate status in pregnant rats results in hypomethylation of the placenta (e.g., Kim et al., 2009). The fact that increasing or decreasing the dietary level of nutrients that act as methyl donors can have direct effects on DNA methylation of placental tissue, and hence on placental gene expression, shows the extent to which external factors can interact with the genetic regulatory machinery to affect fetal development. This is another way in which the placenta becomes a conduit of information from mother to fetus, regulating fetal development via input from the maternal environment.

Our new understanding of genetic regulation, still far from complete, sheds light on why more complex organisms do not necessarily require dramatically larger genomes. Genes, functional regions of DNA that code for one or more proteins, are still important to our understanding of how an organism works, but they are only a small piece of the puzzle. There is far more biochemical information outside of the protein-coding regions than within the protein-coding regions.

Jumping Genes

Transposons, also called mobile DNA or jumping genes, are segments of DNA that can move about the genome within a cell. They are DNA sequences that are cut or copied and then spliced into novel locations in the genome, and they contain repeat sequences at the beginning and end. A large percentage of the genome of humans, and indeed all vertebrates, consists of sequences of repeated DNA, many associated with these jumping genes. An alphabet soup of retrotransposons has been identified in the human genome. The primary division among them is between long terminal repeat (LTR) transposons and non-LTR transposons. Among non-LTR transposons, there are LINEs, short interspersed elements (SINEs), Alu elements, which are a particular class of SINE, and so on. The discovery of the phenomenon of transposition fundamentally changed the perception of the structure of the genetic material underlying inheritance. The genome is more dynamic and mutable than was previously believed.

Two basic classes of transposons exist, called class I and class II and defined by the mechanism through which they are inserted into the genome. Class II transposons (DNA transposons) are cut from the DNA strand by a transposase, which then inserts it into another section of DNA through its enzymatic action. In a modern, word-processing analogy, class II transposons work by cutting and pasting. The transposons discovered by McClintock, *Ac* and *Ds,* are class II transposons. *Ac* is a complete transposon. In other words, the transposon *Ac* contains the DNA that codes for the transposase enzyme, which cuts the transposon from the DNA strand and then cuts another section of DNA and inserts *Ac* at this location. *Ds* has a mutation in its transposase sequence that renders it unable to perform this cut and paste function, but in the presence of *Ac, Ds* is able to "jump" using transposase expressed by *Ac.* Thus is explained the fact that *Ds* only moved about the genome when *Ac* was also present. Miniature inverted-repeat transposable elements (MITEs) are small DNA transposons that consist of short sequences of about 400 base pairs, flanked by inverted repeats of about 15 base pairs. Originally discovered in the rice genome and in *C. elegans*, they have now been found in the human genome. These transposons are too small to encode the necessary enzymes for insertion into DNA; presumably, they are cut and pasted by enzymes produced by larger transposons. In essence they are parasitic transposons, benefiting from the existence of larger transposons that produce functional peptides.

Class I transposons are also called retrotransposons. They differ from class II transposons in that they are not cut from the DNA strand; rather, retrotransposons are initially transcribed into RNA. By the action of a reverse transcriptase, often coded for within the retrotransposon, the RNA strand is reverse-transcribed back into its DNA sequence, and that DNA sequence is then inserted into the genome, often in multiple places. If class II transposons do the biochemical equivalent of cutting and pasting, then class I transposons do the equivalent of copying and pasting. Not surprisingly, retrotransposons are more than ten times as common in the human genome as DNA transposons; retrotransposons and their remnants are estimated to account for more than 40% of the human genome, with DNA transposons accounting for only a few percent (Böhne et al., 2008). Humans are not exceptional among mammals in this regard; roughly one-third to one-half of the mammalian genome is comprised of transposons, a higher proportion than is found in other vertebrates, such as fish (7–12% of the genome) and birds (about 10% of the genome; Böhne et al., 2008). Because many normally silenced retrotransposon sequences are expressed in the placenta, it is an intriguing question to consider which came first, the placenta, or the transposons' contributing a large proportion to the genome.

Retroviruses from the Past

Retrotransposons, the "copy and paste" transposons, exhibit many of the characteristics of retroviruses. Some, perhaps most, retrotransposons are retroviruses that were able to insert themselves into the germline, lost the ability to exit the cell, and now are replicated by their host and transmitted to future generations. Conversely, some retroviruses may have evolved from retrotransposons that acquired the ability to exit cells; a sobering thought—we might be harboring future epidemics within our DNA. The mammalian genome contains many DNA segments that appear to derive from ancient retroviruses. The remains of past infectious agents reside within our DNA. As much as eight percent of the human genome is believed to come from retroviral DNA, about four times greater than the amount of protein-coding DNA. Much of this ancient retroviral DNA appears to no longer be coding, effectively neutralized and rendered a passive traveler on our evolutionary journey. Most of the retrotransposons found in mammalian DNA are silenced, generally by being heavily methylated. Thus, their ability to hop about the genome has been blocked. Doubtless, one adaptive function of DNA methylation may be the silencing of foreign DNA. Methylation of repetitive sequences probably represents a mechanism to defend the genome against infection by retroviruses, to protect the genome after retroviral infection, and to confer resistance to mutational change from endogenous retrotransposons from whatever source. On the other hand, it also serves to maintain the genetic potential represented by the retrotransposon DNA sequences within the genome. The DNA is silenced but not removed from the genome.

Many ancient retroviral sequences are in essence genetic fossils, completely silenced by nonsense mutations and other disruptions of their sequences that have destroyed their function. However, some sequences still contain all the necessary elements to be functional genes and are able to produce functional proteins. Most relevant to this book, there are a number of human endogenous retroviral (HERV) genes that are largely, if not exclusively, expressed in the placenta. Several important proteins produced by the extraembryonic tissues that form the placenta are ancient retroviral proteins. In an exquisite irony, the host has co-opted the genetic machinery of the viruses.

Two examples of co-opted retroviral genes expressed in the human placenta are syncytin 1 and syncytin 2. These envelope-protein genes from HERVs enable cytotrophoblast cells to fuse and become the multinucleated syncytiotrophoblast. Syncytin 1 is an envelope protein from ancient human endogenous retrovirus HERV-W. Syncytin 2 does not appear to be a replication of syncytin 1 but rather derived from an independent later retroviral infection, HERV-FRD. Although it has fusogenic properties similar to syncytin 1, syncytin 2 also has immunosuppressive properties. These genes are found in all anthropoid primates and are remarkably conserved, indicating strong selection for their function (Blaise et al., 2003). They are not found in prosimian primates, indicating an origin of between 40 and 70 million years ago.

This phenomenon is not unique to anthropoid primates. Murid rodents possess related genes, termed syncytin A and syncytin B, that perform the same function. Interestingly, these retroviral genes are not orthologs of each other. In other words, these four syncytin genes are not related by common descent, at least not in a simple manner. Their ancestor was not present in the common ancestor of primates and rodents. Instead, they appear to have independently invaded the germlines of anthropoids and murid rodents via four different retroviral infections. This is an example of horizontal, rather than vertical, gene transfer—new DNA coming into species from outside.

In placental tissues, many other retroviral genes are active, with various functions, most not yet understood. Placental tissue is relatively hypomethylated, with significantly fewer base pairs methylated compared to other cells, especially within repetitive sequences (Ng et al., 2010). This hypomethylation is not random and allows more of these usually silenced genes to be expressed. Disruption of DNA methylation in rodent models either by exogenous administration of demethylating agents (Vlahovic et al., 1999; Šerman et al., 2007) or by knockout studies of DNA methyltransferase genes is associated with placental defects (Li et al., 1992; Arima et al., 2006).

These facts lead to some interesting and unresolved questions. To what extent do certain placental mammal lineages owe their evolutionary success to ancient viral diseases? How have retroviruses contributed to the diversification of placental biology within the placental mammals? Has the placenta made mammals more susceptible to retroviral infections invading the germ line?

Retroviral Remnants as Genetic Fossils

The remnants of retroviral infections that invaded the germline and have then been neutralized by the various gene-silencing mechanisms can serve as a genetic fossil record. These DNA sequences are handed down in a Mendelian fashion, just as any other locus. If two species have remnants of a retrovirus at syntenic chromosomal locations (i.e., they are located on chromosomes that have a common origin), that is strong evidence that the original infection occurred prior to those species having diverged. In other words, the infection occurred in a common ancestor. Comparing the two species, the extent of difference in DNA sequence of the retrovirus genes is a measure of how long ago they diverged from each other. In contrast, if species B does not share syntenic retroviral DNA found in species A, that is strong evidence that the infection occurred after they diverged. If the time the infection occurred in species A can be estimated, then this gives a minimal time for the divergence between the two species. Thus, retrotransposons and especially their long terminal repeats can be useful tools for determining phylogenetic relationships and for dating divergences.

Incorporation of retroviral genes into a species is not just a theoretical construct that has led to consistent predictive findings when related species are examined. Unfortunately, in the koala, scientists are witnessing this phenomenon in the present (box 6.1).

Genetic Regulation

Gene regulation is the key to producing complex organisms. The genes that code for proteins are of course important, but the complexity of life comes from the mechanisms by which gene expression is regulated. This has long been known. Ever since DNA was discovered and confirmed to be the (primary) source of heritable genetic information, scientists have wrestled with the contradictory and opposing notions that the genetic coding passed on from parent to offspring has to be both relatively invariant (representative of the parents) and have the capacity to produce radically different cells. The ge-

BOX 6.1 KOALA RETROVIRUS

That Australian icon, the koala, unfortunately appears to be undergoing a germline invasion by a retrovirus, the so-called koala retrovirus (KoRV). The virus is ubiquitous in Queensland, with 100% of animals testing positive (Tarlington et al., 2006). The virus is not found in the south of Australia. In the early 1900s, an island off the south coast of Australia was stocked with koalas, which have been largely isolated from the mainland populations ever since (Stoye, 2006). These animals have tested negative for KoRV.

Viral sequences have been shown to be present in sperm, suggesting an endogenous retrovirus. However, the extent of the variation in the KoRV envelope gene and the existence of a variable copy number of the virus among individuals in the infected population is consistent with an exogenous virus (Stoye, 2006). The existence of uninfected animals in geographically separated habitats also implies that KoRV is an active exogenous retrovirus that has invaded the germline in some populations and is now being spread by both infection and Mendelian inheritance. If the koala survives this threat to its existence (KoRV is linked to a fatal lymphoma), future generations of koalas are likely to carry an endogenous retrovirus descended from KoRV within their genome.

netic program passed on through the generations is far more than a list of transcribable DNA segments that produce RNA that serves to assemble proteins. Somehow, information on the timing and relative expression of gene products is also passed down. The modern model of a gene contains regulatory segments. Genes have promoters that regulate transcription. Promoters often are activated by molecules binding to them or turned off by other molecules binding. The categories of molecules that bind to segments of DNA to influence DNA transcription continue to grow.

There are time scale differences in gene regulation. Short-term regulatory processes serve to drive physiological adaptation to the environment. Ligands bind to receptors, which then bind to promoter regions of DNA to affect transcription. Longer-term mechanisms act as well. In cell differentiation, genes are turned on and off, and they remain on or off for variable amounts of time. In addition, instead of simply "on" or "off," their expression can be regulated to be high or low. This can be accomplished by regulation of transcription at many different points in the process, and it can also be accomplished by variable on-and-off switching among cells. In other words, not every cell in an organ needs to express its genes identically; the pattern of expression can be restricted. And finally, biochemical processes can act both post-transcriptionally, affecting the RNA, and post-translationally, affecting the peptide.

For the purposes of this book, we divide gene regulation into two broad categories. One is the ongoing regulation of genes and gene products that in turn regulate metabolism and physiology. Genes are turned on and off, upregulated and downregulated, in response to external and internal signals. The regulation is transient and ever changing. In one set of circumstances a gene will be switched off; some time later—perhaps a long time, perhaps quite a short time—the gene will be switched back on. This is the dynamic regulation that makes life possible.

The other broad category corresponds to cellular differentiation. Pluripotent cells, the embryonic stem cells that have been so much in the news, are capable of becoming any cell type. But an organism is not made of pluripotent cells; rather, an organism consists of coherent groups of different cell types that function together to form organs and systems. These different sets of cells have different sets of genes that are activated and functioning. In essence, cell lineages become channeled down developmental pathways, with sets of genes turned on or off, upregulated or downregulated; for those genes the expression levels and dynamic regulatory settings (the first category of gene regulation) will be at a level characteristic for that cell type. This regulation is generally more static (nothing in life is completely static) in that the genes that are set to be on or off rarely, if ever, change. This represents a much more permanent form of regulatory control and generally relies on different mechanisms to accomplish genetic regulation.

The placenta arose from genetic potential that is quite ancient. The extraembryonic membranes of the placenta derive from the membranes of the amniote egg (yolk sac, amnion, chorion, and allantois). Of note, there are relatively few genes that appear to be exclusively expressed by the placenta, and those vary significantly among species (Rawn and Cross, 2008). The development of the placenta relies to a large extent on: co-opted function of existing genes; gene family duplications, where some gene replicates are expressed predominantly in the placenta; changes to regulatory regions of the genome that change gene expression in placental tissues; and, amazingly, co-opted retroviral genes from long-ago infections. Methylation of DNA is significantly less common in placental tissue compared to most other tissues (Fuke et al., 2004), providing a mechanism for gene expression by normally silenced genes, such as retroviral genes. The modern understanding of genetics emphasizes regulation and variability in gene expression and in gene product formation and function. The placenta provides many examples of the complexity and flexibility of the genetic system.

Cytogenetics to Genetics and Back

The earliest investigations of genetic material centered on the structure of chromosomes. With the discovery of DNA and later its association with RNA, the focus of genetics switched to genes, DNA sequences, transcription, and translation into peptides. The sequencing of whole genomes was thought to be the final step in understanding how complex organisms were constructed from the instructions inherent in genetic material. Biologists were somewhat surprised when the sequencing of the human genome discovered a mere thirty thousand, or possibly fewer, protein-coding genes and not the hundred thousand that had been predicted. Our genome appeared to be bigger than needed for that number of genes, and the number of genes seemed to be smaller than expected to produce an organism as complex as we like to think we are. Perhaps ironically, biology is now somewhat retracing its steps back to cytogenetics in investigating the structure of the packaging of the DNA as a mechanism for genetic regulation. Two mechanisms of gene silencing (histone modifications and DNA methylation) both involve modifications of DNA structure that affect DNA transcription without changing the underlying DNA sequence. Histone modifications in general result in less-permanent gene silencing than does DNA methylation. The two mechanisms have been shown to interact, and some have suggested that histone modification represents an older mechanism. These modifications have been seen to enhance gene expression as well as silence it, either indirectly, by modifying the expression of other regulatory DNA sequences, or perhaps even directly, by enhancing rather than suppressing transcription. The mechanisms by which these epigenetic processes function are not fully understood. Nonetheless, these epigenetic mechanisms that regulate gene expression by changing the structure of the genetic material, but not its sequence, are important aspects of the overall developmental plan that creates complex organisms.

Assisted Reproductive Technology and Epigenetics

The first human baby conceived by in-vitro fertilization (IVF) was born in 1978. The proportion of babies born using some form of assisted reproductive technology (ART) is small though increasing, but even a small percentage results in a fairly large absolute number of babies born using ART. The vast majority of these individuals appear developmentally normal. However, since the oldest are only in their thirties, the long-term outlook for adult-developing diseases such as cancer is still uncertain. Large-sample studies of children conceived through ART have documented an increase in congenital malformations over what was expected as well as an increased risk for certain disorders such as Angelman syndrome and Beckwith-Wiedemann syndrome (Wilkins-Haug, 2008).

Both Angelman syndrome and Beckwith-Wiedemann syndrome can be caused by loss of methylation, leading to expression of, usually, maternally-imprinted genes (Wilkins-Haug, 2008). One aspect of the mammalian system of genome regulation relies on methyl transferases to regulate DNA methylation, with histone modifications and other epigenetic marks also serving to guide DNA methylation. How these epigenetic marks are maintained, removed, or reapplied is not fully understood. There is concern that the conditions of IVF may influence DNA methylation. This will affect gene expression; most concerning, it might affect gene imprinting. Environmental influences affect development down to the genome level; this is an evolved capacity that enables maternal circumstances to modulate fetal growth and development, presumably to, on average, produce offspring better adapted to the circumstances they will be born into. Our development of technologies to assist in reproduction may produce unintended consequences via these mechanisms, providing a practical motivation for studying these genetic and epigenetic regulatory mechanisms.

Early Placental and Embryonic Development

The placenta is genetically identical to the fetus, but obviously the placenta takes a vastly different developmental path. Of course, within the fetus cells become differentiated and take various developmental routes. Identical genetics expressed differentially serves to create a complex, multi-organ individual. How cells become differentiated has long been a major question in biology. Modern science is on the cusp of being able to accurately describe the processes and mechanisms by which it happens. Our conception of the genome has changed from a (relatively) static blueprint for growth and development to that of a dynamic, flexible organizer and regulator of information that enables appropriately adaptive growth and development under diverse conditions.

The cells that are to form the mammalian placenta are identifiable very early post fertilization. Briefly, after a placental mammal ovum is fertilized by a sperm (forming a zygote) the now-diploid cell reenters the cell cycle and begins to divide. Cell division continues without any increase in the total size of the cell mass, a process called cleavage. The cells (at this stage called a blastomere) form a tightly massed ball enclosed within the zona pellucida, a glycoprotein structure that surrounds the ovum. The zona pellucida is retained until just before implantation. In this phase, the embryo is termed a morula. The zona pellucida is lysed (zona hatching) and the embryo can then implant.

The next phase is blastulation. The blastomere becomes divided into polarized and apolar cells, with the polarized cells on the outside. The cells in the outer cell mass pump sodium ions into the center of the cell mass, which brings in water through osmosis. The embryo is now at the blastula stage, with an outer cell mass and an inner cell mass pushed to one side of a fluid-filled cavity. The outer cell mass becomes trophoblast and will form the chorion of the placenta. The inner cell mass will produce the fetus and other extraembryonic membranes (the amnion, which will form the amniotic sac, and the allantois, which will become the umbilical cord.) So, by the blastula stage (within the first week post conception in humans) there are cells (the outer cell mass) destined to become the outer membrane of the placenta (chorion) and cells in the inner cell mass that will form the embryo, the amniotic sac that protects the embryo, and the allantois, which will eventually fuse with the chorion to enhance transport to and from the embryo.

This cellular arrangement is unique to placental mammals. Marsupial embryos do not have inner and outer cell masses; rather, the trophoblast (placental) and epiblast (embryonic) cells occupy opposite ends of the blastocyst. Certain marsupial embryonic cells will form a placenta and others will form the fetus, but they are not arranged in the multilayered fashion that is unique to extant eutherian mammals.

Placental Gene Expression

There appears to be an interesting pattern to the timing of placental gene expression across gestation. Fascinating research by Kirstin Knox and Julie Baker of Stanford has shown that gene expression in the mouse placenta follows a pattern of early expression of ancient genes common to most vertebrates, followed by later expression of more recently evolved, rodent-specific genes (Knox and Baker, 2008). In summary, they found that gene expression in the developing mouse placenta at embryonic day 10.5 (e10.5) or before was enriched in orthologs of truly ancient genes, genes that are considered orthologs of eukaryote gene ancestors. In the mature mouse placenta (e15 and beyond) there was no enhancement with ancient gene expression; instead, a significant increase in expression of rodent-specific genes was seen. This change in gene expression occurs despite the similarity in physical form between the mouse placenta at e10.5 and the mature placenta and cannot be explained by the size difference or an increase in the percentage of a particular cell type. Rather, there appeared to be a global change in gene expression over time. A similar examination of gene expression in term, human placentas found enrichment in expression of primate-specific genes.

A survey of the ancient genes that are highly expressed in the early mouse placenta is instructive. They are predominantly genes related to growth and metabolism, including cellular metabolism, cell division, and the cell cycle (Knox and Baker, 2008). In other words, genes that are relevant for rapid growth of the extraembryonic tissues would appear to be upregulated. Assuming that what happens in the mouse placenta is representative of the pattern of gene expression change in placentas from other species, it would appear that during initial development the genes that are expressed in the placenta are ancient genes, conserved across placental mammals; not surprisingly, however, as development progresses, the gene expression patterns begin to diverge, such that human, rat, and mouse placental gene expression become progressively more different.

Evolutionary Descent of Placental Genes

In most instances, the genes that are expressed in the placenta are the same as those expressed in the mother and the fetus; the number of placenta-specific genes appears to be fairly low (Rawn and Cross, 2008). Most hormones, for example, are ancient and predate the evolution of placental mammals, so these molecules would not be expected to come from placenta-specific genes. For most genes in the placenta, the dissimilar functional outcomes are explained by differential regulation of the genes in the various tissues.

Placental-specific gene function has occurred due to a set of mechanisms. In addition to changes in epigenetic marks, as discussed above, novel promoter regions have evolved that modify transcription of genes. Novel post-translational modifications of gene products produce placenta-specific functions. Also, there have been gene duplications, for instance, of the growth hormone gene in anthropoid primates and the prolactin gene in rodents and ruminants. The evolution of the growth hormone / prolactin gene family is an example of how evolution has resulted in expansion of an existing gene into a related gene family, producing multiple signaling/information molecules that have acquired new functions in the placenta.

The GH/PRL Gene Family

A family of polypeptide hormones with lactogenic and somatogenic properties are expressed by rodent, ruminant, and anthropoid primate placentas. This peptide family is composed of the placental growth-hormone variants, prolactin, prolactin-related peptides, and the placental lactogens. These peptides are all structurally related and are thought to derive from a single common-ancestor gene that existed in the vertebrate ancestor more than 345 million years ago. A number of different successive duplication events in disparate lineages have produced the various gene families found in extant vertebrate species. The first duplication in the lineage leading to all vertebrates resulted in prolactin (PRL) and growth hormone (GH). The original functions of PRL and GH may have been related to osmoregulation. Certainly these hormones have important osmoregulatory effects in fish (box 6.2). Subsequently, independent duplications of these two genes occurred in several placental mammal lineages, which resulted in the placental lactogens (PL). In general, these hormones are glycoproteins. In other words, they are peptides with attached sugar residues, meaning that they have been glycosylated after transcription.

The first evidence for an extended PRL family of bioactive peptides expressed by the placenta in rats and mice came from studies in the 1930s examining the effects of removing the anterior pituitary gland during pregnancy in these rodents (e.g., Pencharz and Long, 1931; Selye et al., 1933; Astwood and Greep, 1938). Prolactin secreted from the pituitary was believed to be necessary for the continued function of the corpus luteum and the development of the mammary glands. Yet removing the pituitary after mid-gestation did not disrupt these processes, suggesting that another organ (i.e., the placenta) produced either PRL or peptides with PRL function. And, indeed, placental extracts were shown to have luteotrophic and lactogenic activity (Deanesly and Newton, 1940; Cerruti and Lyons, 1960).

Rodents (e.g., mice and rats), ruminants (e.g., cows and sheep) and anthropoid primates (but not prosimian primates) also engage in placental production of bioactive molecules with prolactin-like activity. Interestingly, the anthropoid placental lactogens appear to derive from a duplication of GH, while in rodents and ruminants the PL peptides are derived from PRL (Soares et al., 1998). In rats, the expanded PRL family encompasses about 1.7 Mb (mega base pairs) on chromosome 17. In mice, it is composed of about 1 Mb on chromosome 13, and in cows it is on chromosome 22. These regions are syntenic with each other and with PRL loci on human chromosome 6 and dog chromosome 35. The evidence strongly favors a series of gene duplications, starting with a duplication of the PRL gene, to explain this distribution of PL genes in rodents and ruminants.

Humans and other anthropoid primates also express placental lactogens. However, the PL genes in humans are not on chromosome 6 but on chromosome 17, in proximity to the GH gene (Soares et al., 1998). Thus, the evolution of placental lactogens in primates and rodents is an example of parallel evolution, in which different (but in this case related) genes are duplicated, and the duplicated genes have undergone selective change to acquire an extended range of functions. Rodents and anthropoid primates have independently solved similar adaptive challenges through a similar process (gene duplication), using different (but related) genes.

The human GH locus consists of five genes. One (GH1) is expressed by the pituitary; the other four (GH2, CSH1, CSH2, and CSHL1) are expressed by the placenta. The four placental genes are derived from duplications of GH and additional promoter regions. All of the placental GH-family genes have multiple splice variants, four each for GH2 and CSHL1, and three each for CSH1 and CSH2. The pituitary-expressed GH displays signs of purifying selection constraining its sequence, while the placentally-expressed variants display evidence of strong positive selection (Ye et al., 2005; Papper et al., 2009).

Other anthropoid primates also have a GH cluster of genes. Rhesus macaques have a gene cluster similar to that seen in humans, with expression of the four duplications in the placenta (Golos et al., 1993). The common marmoset and the capuchin, two New World monkeys, have variable numbers of GH genes and pseudogenes. Eight GH-related genes have been identified in the marmoset, while the capuchin GH cluster has at least 40 GH-related genes, although most are probably pseudogenes (Wallis and Wallis, 2006). Four GH genes have been identified in the spider monkey, with three expressed in the placenta and one only in the pituitary. A comparison of the GH-family gene sequences for Old and New World primates suggests that a single GH variant existed in their last common ancestor and that the two gene clusters are a re-

BOX 6.2 **GH/PRL AND OSMOREGULATION**

Fluid balance and osmoregulation are vital processes in living organisms. Many of the ancient vertebrate hormone families may have been derived from ancestor genes that were involved in these critical processes. This would appear to be true for the GH/PRL family, based on the evidence for their function in teleost fish. In these fish, both GH and PRL are used in osmoregulation (Sakamoto and McCormick, 2006).

All teleost fish maintain their extracellular fluid at about one-third the osmotic value of modern seawater. Teleost fish live in waters with a range of ionic concentrations, from freshwater with very low osmotic values, to salty seawater. Many species (e.g., trout and salmon) move back and forth between saltwater and freshwater. So teleost fish essentially find themselves always either hyposmotic or hyperosmotic to their environment. Fish in seawater face the adaptive challenges of being dehydrated and salt loaded, while freshwater fish face overhydration and salt depletion.

In the 1930s, the physiologists Homer Smith, August Krogh, and Krogh's post-doctoral assistant, Ancel Keys, undertook a series of studies to explore how fish (and other water-dwelling organisms) solved these osmotic challenges. Freshwater fish produce copious quantities of dilute urine to counteract the substantial influx of water into their tissues through diffusion. Marine fish produce small volumes of urine that is osmotically similar to seawater. But these adaptations do not solve the problem by themselves. Freshwater fish still face a challenge in replacing the salts that

continued

sult of independent duplications in the two lineages (Papper et al., 2009). This implies that the original anthropoid GH was expressed in both the pituitary and the placenta. In anthropoid primates, pituitary GH can stimulate the PRL receptor; this is not true for rodent or bovine GH. Thus, in the anthropoid primate lineage, GH and GH-related genes could be co-opted into serving placental lactogenic functions. Duplications of the GH gene to form a GH-family locus of expressed genes likely would not have resulted in placental lactogens in rodents.

Chorionic Gonadotropin

There are two important placental signaling molecules in anthropoid primates that appear to be unique to that lineage: corticotropin-releasing hormone and chorionic gonadotropin. CRH is an example of an existing gene with impor-

will be lost to their environment. In seawater, the adaptive difficulty is to replace the water lost to diffusion from extracellular fluid into saltwater, while regulating ionic balance, since salts will be drawn into the body.

How are these challenges met? Marine fish drink seawater and actively absorb salt (NaCl) and water across the gut.The absorbed sodium chloride helps transport water across the gut epithelium by creating an osmotic gradient. The water moves with the sodium. The sodium and chloride ions are then excreted independently of each other by the gills. The chloride ions are excreted by special cells (called chloride cells) that have high densities of mitochondria, which enable them to metabolize the necessary energy to pump ions against the osmotic gradient. The sodium ions are excreted via gap junctions associated with the chloride cells. In freshwater, chloride cells are modified to be able to reverse the process, absorbing Na^+ and Cl^- ions, even in water with very low concentrations of ions (Sakamoto and McCormick, 2006).

A number of types of fish are faced with adapting to both seawater and freshwater, either during different parts of their life cycle or, for some, on a shorter time scale, due to living in tidal areas at the mouths of rivers, where the salinity of the water can change dramatically because of tides or changes in river flow. Both PRL and GH have important functions in the necessary changes fish must undergo when they move between seawater and freshwater. In going from salty water to freshwater, PRL and GH act to regulate the reorganization of morphology and physiology of the gills to change them from organs that excrete excess ions to ones that absorb these ions from the new, low-salinity environment.

tant functions throughout the body that has been co-opted in the anthropoid placenta for a unique endocrine circuit linking maternal, placental, and fetal physiology. The only change at the genome level was to allow expression of CRH in placental tissue and secretion into both the maternal and fetal compartments. We discuss CRH in greater detail in chapter 7.

The other signaling molecule, CG, is a novel peptide created by a duplication of one of the units of the luteinizing hormone (LH) gene. Notably, a peptide with CG-like activity is also found in equids, as mentioned in chapter 3, where we discussed the horse epitheliochorial placenta. This would appear to be a case of either convergent or parallel evolution.

CG is a member of a family of heterodimeric cysteine-knot glycoproteins whose members have a structure composed of a common α unit and a unique β unit. The α unit for CG is identical to the α unit for thyroid stimulating hormone (TSH), leuteinizing hormone (LH), and follicle-stimulating hormone (FSH). All of these hormones have a unique β unit that produces the particular biological function for each molecule. The β unit for CG is derived from a gene duplication of the β unit for LH. A broad survey of the genomes of primates and their near relatives detected variable numbers of CG β-unit genes in anthropoid primates but none in tarsiers, lemurs, lorises, or any nonprimates (Maston and Ruvolo, 2002). In New World monkeys, only a single CG β-unit gene has been found, but in Old World monkeys and apes, the number ranges from three in macaques to five in chimpanzees and dusky leaf monkeys (Maston and Ruvolo, 2002; Hallast et al., 2008). Humans are the champs, with six CG genes identified (Policastro et al., 1986).

The instance of the convergent phenotypic evolution in equids is quite different; there was no gene duplication. Equine LH and CG (eCG) are coded for by the same two genes: the common α unit and the equine LH β-unit gene (Sherman et al., 1992). The difference in function is accomplished by varying glycosylation events at the terminal end of the peptide, with eCG being more glycosylated. In other words, in the horse, post-translational biochemistry results in two molecules with different functions, another example of how there can be many solutions to the same biological challenge.

In addition to the original gene duplication from which the CG β-unit gene arose, a number of other events occurred to create this novel functional peptide in human placental physiology. A frameshift mutation occurred in the CG β-unit gene that allowed translation to continue into the LH gene 3' untranslated region, adding 24 amino acids to the transcribed peptide. This addition of amino acids increased the susceptibility of the peptide to glycosylation, resulting in a molecular weight much greater than that of LH as well as a longer half-life in circulation, 8–16 hours compared to 20–40 minutes for LH (Henke et al., 2007). Expression of the CG β-unit gene was enabled in the placenta and, at some point, suppressed in the pituitary; and of course, expression of the common α unit gene had to be enabled in the placenta. Why there were subsequent additional duplications of the CG β-unit gene in Old World anthropoids is uncertain. In New World monkeys, a quite different outcome occurred. In at least three species of New World monkeys (common marmosets, squirrel monkeys, and owl monkeys) LH has become a pseudogene, and its function in the pituitary has been assumed by CG (Henke et al., 2007; Scammell et al., 2008). In chapter 7, we discuss the functions of human CG (hCG) and differences in expression and function among the six different β units.

Some authors have proposed that CG was an important component of the evolution of a hemochorial placenta in anthropoid primates and that its importance in human implantation and gestation has been increased, in association with the interstitial implantation and deeply invasive placental phenotype of human beings (e.g., Cole, 2009). However, the best evidence suggests that a hemochorial placenta is the ancestral condition in primates (Martin, 2003; Wildman et al., 2006). The closest living relative to primates, the colugo, has a hemochorial placenta but only expresses LH, with no evidence of any CG genes (Maston and Ruvolo, 2002). Thus, it would appear that a hemochorial placenta, due to the increased access to the mother via the direct connection with maternal blood, may have allowed this gene duplication to achieve adaptive function, but CG was not involved in the early evolution of a hemochorial placenta in anthropoids. Horses, plainly, have solved the challenge of placental CG signaling in a completely different way, producing transient endometrial cups early in pregnancy that erode into maternal epithelium and secrete CG into the maternal circulation.

Whether the evolution of hCG has any association with human implantation characteristics is an intriguing question. The evidence does indicate that there is an extra CG gene in the human lineage (six, versus five in chimpanzees; Hallast et al., 2008). Three of these genes (hCG 5, 3, and 8) produce identical proteins (Miller-Lindholm et al., 1997). Gene hCG 7 differs from the other three genes in only a single amino acid substitution (Policastro et al., 1986). The hCG genes 1 and 2 may produce proteins with different functional effects, as they have a changed open reading frame from the other genes. Among both normal and abnormal placentas, hCG genes 5, 3, and 8 are the most commonly expressed, followed by hCG gene 7 (Miller-Lindholm et al., 1997). There is a large amount of variability in which particular gene is expressed in placentas, though gene 5 is generally the most commonly expressed; neither gestational age nor any other placental characteristics appear to explain this variation in expression. It would appear that the total amount of hCG expressed is vital for successful pregnancy (Miller-Lindholm et al., 1997), but exactly which hCG genes are translated does not appear to be critical.

As we discuss in more detail in chapter 7, a form of CG (hyperglycosylated CG) assists in regulating the depth and extent of trophoblast invasion and remodeling of the maternal spiral arteries. Although the duplication of the LH β gene was unlikely to have been associated with the development of a hemochorial placenta in anthropoid primates, the existence of CG may have been involved in the evolution of interstitial implantation with a deep invasion of decidual tissue. It is as probable that CG was co-opted into a regulatory function regarding the depth of trophoblast invasion. In the next chapter we discuss the regulatory nature of the placenta on many levels, including the various roles hCG plays in successful gestation.

7

The Placenta as a Regulatory Organ

Many kinds of regulatory systems exist, both within and outside of biology. An important distinction among them is that biological regulatory systems have evolved, and the process of evolution endows them with certain aspects that may not be present in a designed regulatory system. Frequently, redundancy of signaling and function occur. Often, different molecules will act through disparate receptors but still activate the same downstream cascades (or ones that have similar effects). In addition, often the same signaling system has a diversity of function. Different molecules can act through the same receptor but initiate dissimilar downstream cascades, or the same molecule and receptor can interact to cause disparate effects at diverse developmental stages or in different tissues, due to variations in intervening molecular pathways. In many cases there has been duplication of genes, resulting in a gene family whose various members may have similar or quite different functions.

The placenta serves as a central regulator of maternal-placental-fetal metabolism (Power and Schulkin, 2005). Regulatory molecules pass from maternal circulation to the fetus via the placenta, and molecules produced by fetal organs end up in maternal circulation. More importantly, the placenta is a potent synthesizer and secretor of a wide array of information molecules that act on maternal and fetal metabolism and development as well as on the placenta itself (Petraglia et al., 2005). The placenta coordinates maternal physiology with fetal development and even has a role in stimulating maternal changes that prime the mother for caregiving behaviors after birth (Numan et al., 2006). The human placenta is fascinating—a transient endocrine organ that regulates and coordinates metabolism and physiology among the mother, the fetus, and itself.

Regulatory Physiology

Some argue that the science of regulatory physiology began with the work of Claude Bernard in the 1800s. The concepts of the internal milieu itself, and also the importance of that milieu remaining constant when challenged by external environmental perturbations, were essential breakthroughs in understanding regulatory physiology. Other scientists, such as Walter Cannon, extended and refined Bernard's concept to arrive at the paradigm of homeostasis. Homeostasis is a simple but powerful concept. The external environment not only continually changes, but even in periods of stability, it is rarely conducive to the necessary chemical reactions (metabolism) necessary for vertebrate life. Thus, vertebrates must have mechanisms to maintain an internal chemical state that can support metabolism, even in the face of external variability. The internal milieu must remain constant, despite the external challenges. Otherwise, the animal dies. In the words of Cannon (1935), "the organs and tissues are set in a fluid matrix . . . [S]o long as this personal, individual sack of salty water, in which each one of us lives and moves and has his being, is protected from change, we are freed from serious peril."

The paradigm of homeostasis has been particularly fruitful for understanding much of regulatory physiology, but not all. Many challenges animals face demand change rather than stability. Stability is not the currency of evolutionary success; reproductive fitness is the better measurement. Physiological regulation does not act to maintain stability but instead to maintain viability. In many instances, that means abandoning stability and embracing change in order to survive and reproduce. The concept of allostasis, defined as changing an internal state in support of viability, has been added to homeostasis, which is defined as maintaining an internal state in support of viability.

Pregnancy is an excellent example of where homeostasis appears to fail, and the concept of allostasis may be more appropriate. Female metabolism undergoes a series of adaptive changes during pregnancy. Pregnancy requires a relaxation of homeostasis, in part because metabolism and physiology must account for two organisms instead of one. Many physiological set points must be adjusted, often many times, during gestation. For example, circulating levels of glucose, insulin, cortisol, and leptin all increase. Pregnant women are termed hyperglycemic, hypercortisolemic, hyperinsulemic, and hyperleptonemic. From this terminology, one might be forgiven for thinking pregnancy is a metabolic disease. But these terms just relate the adaptive, pregnant condition to the more homeostatic nonpregnant state. Maternal insulin resistance during pregnancy is not pathological, at least as long as it remains within the viable limits that characterize the pregnant metabolic phenotype. Indeed, maternal insulin resistance can be considered an example of allostatic regulation, in which physiology is continually adapting to changing circumstances in order to maintain viability. In evolutionary terms, viability requires passing genes on to the next generation. Pregnancy obviously is the main female mode for achieving reproductive success in mammals. It is not surprising that understanding pregnant metabolism requires more than an understanding of the strategies for survival.

Yet, because of the fundamental mammalian adaptation for reproductive success (lactation), maternal survival must be a part of the mammalian reproductive strategy. There are evolutionarily successful species in which individuals (both male and female) die after reproduction (e.g., cuttlefish, salmon.) That doesn't work for a mammal. A female mammal that dies after parturition is not a successful female. In general, mammalian young require substantial postpartum care. The closest a mammal comes to the "produce them and leave them to their own devices" reproductive strategy is when, in some marine mammal species, for example, the female hooded seal gives birth to a pup on unstable pack ice, nurses it for just under four days (during which time the pup doubles in size) and then leaves, for all we know never to see her pup again. This is highly unusual among mammals, though it demonstrates the breadth of reproductive strategies potentially available to mammals, despite the constraints of gestation and lactation. Indeed, the existence of a placenta enables the production of such large and precocious offspring that lactation can be abbreviated. But even in this case, the female must survive giving birth. Human beings are long-lived, slowly reproducing animals. Our female ancestors produced multiple offspring spread out over tens of years. Thus, whatever physiological adjustments must be made during gestation should not be so extreme as to pose a long-term threat to maternal health.

Other terms have been suggested for the altered regulation of metabolism to support the demands of gestation (and lactation); for example, Bauman and Curie (1980) suggested homeorhesis. The exact term is not as important as the concept that gestational metabolism cannot fit the homeostatic model in all cases. Certainly, maternal physiology during gestation will be homeostatic in many aspects, but in others it must be constantly changing in order to remain adaptive.

Regulation is the key to survival. Animals are constantly adjusting their physiology and metabolism in order to remain within the bounds of viability. The new and exciting understanding in genetics is that, at the level of DNA, regulation is also the key to viability. The old thinking was that after sequencing the genome, we would understand the workings of life. The notion was that the sequences of DNA would largely determine the growth, development, and metabolism of an animal. Although in essence that is true, we now understand that the complexity of the system is greatly enhanced by regulatory elements, modifications to the DNA, and post-transcriptional modifications and metabolism of gene products. Transcription of DNA to assemble a functional protein is a necessary function of the genome but not the most important for understanding how a complex organism functions. Our new comprehension of genomics brings genetic regulation closer to our conception of regulatory physiology. Metabolism occurs at all levels in an organism: the organism level, the organ level, the cellular level, and the genome level.

Information Molecules

The placenta is an extra-fetal organ that produces as many as, or more of, the molecules of life than any other organ. Although all organs contain the genetic potential of the whole organism, various regulatory mechanisms result in unique patterns of gene expression for every organ. And of course the placenta is not uniform—it contains various structures and tissue types. Of central importance, the placenta has two sides: one that connects to the fetus and one that connects to maternal tissue. The placenta can send signals to both the mother and the fetus. It is the conduit for the information that passes between these two separate-but-connected organisms, but in many cases it is the originator of the signals.

The molecules of life can be classified in many different ways. Some are called nutrients, some are called enzymes, or hormones, or cytokines. Some molecules have structural functions, others function in metabolic processes, some in information signaling and regulation, and some in immune function. Many of these molecules have multiple purposes across several categories. For example, many cytokines can have an immune function in one context and a regulatory function in another. As we discuss in more detail later in this chapter, during human implantation the interaction between trophoblasts and the maternal immune system regulates maternal tissue remodeling and the depth of trophoblast invasion. The immune system acts as a regulatory system for the growth and development of tissue.

Placental Growth Hormone

In chapter 6, we discussed the human growth hormone family, which consists of five genes, one expressed in the pituitary (GH1) and the other four in the placenta. The expression of pituitary growth hormone and the placental variant (GH2) during pregnancy is a classic example of the placenta acting as a central regulator of maternal physiology.

GH1 is synthesized and released into circulation by the anterior pituitary, under the control of a well-described circuit that regulates GH1 production and release. Growth hormone–releasing hormone secreted by the hypothalamus induces GH1 synthesis and secretion by the pituitary; somatostatin is also released by the paraventricular nucleus (PVN) of the hypothalamus and acts to downregulate GH1 synthesis and release. Interestingly, ghrelin, a gut peptide linked with appetite and the organization of feeding behavior, can also stimulate GH1 secretion (Kojima et al., 1999).

During human pregnancy, this circuit is altered; starting about the twenty-fourth week of gestation, GH1 hormone synthesis and secretion from the maternal pituitary declines, until it essentially shuts down after the second trimester. A placental growth hormone (GH2) is secreted by the placenta and effectively replaces pituitary GH1 in maternal physiology. The synthesis and secretion of GH2 by syncytiotrophoblast begins in mid-gestation. Placental GH2 is not regulated by GH1-releasing factors but is suppressed by elevated maternal glucose. The two hormones are coded for by separate genes from gene duplication events (see chapter 6). They have different receptor binding function, with GH1 binding to both growth hormone receptor (GHR) and the prolactin receptor (PRLR), while GH2 binds to GHR but has little affinity for PRLR. The three other placentally expressed GH-family genes have lactogenic activity, acting through PRLR. The range of functions of GH2 is not completely clear, but it likely serves to induce relative maternal insulin resistance and encourages reliance on lipolysis for maternal energy metabolism, sparing glucose for maternal brain and fetal/placental metabolism. Maternal insulin-like growth factor 1 (IGF-1) circulating levels are correlated with circulating GH2 levels, suggesting that GH2 regulates maternal IGF-1 production. Thus, in this instance the placenta performs a role in the regulation of maternal physiology that before pregnancy was coordinated by the central nervous system. After parturition, circulating GH2 levels rapidly fall to zero, and the pituitary must resume its GH1-secretion function.

The evidence for placental production of growth hormone in rodents is uncertain. As discussed in chapter 6, in rodents it would appear that duplications of the prolactin gene, rather than the growth hormone gene, have created novel peptides with important roles in placental functions. However, there are two growth hormone genes in the ovines (sheep and goats), and sheep express growth hormone from the placenta. In sheep, secretion of growth hormone is into the fetal compartment. The existing evidence suggests it is unlikely that any significant secretion into the ovine maternal compartment occurs. In humans, evidence of secretion into the maternal compartment has been found, but placental growth hormone has not been located in fetal blood. It is found in amniotic fluid, however, and thus must enter the fetal compartment somehow. Levels of GH2 in amniotic fluid were higher mid-gestation than at term and were not affected by labor (Mittal et al., 2008). Increased levels of GH2 are noted in the amniotic fluid of Down syndrome pregnancies (Sifakis et al., 2009). The functions of GH2 in the fetus are not well understood; since amniotic fluid is swallowed, it may have effects on gut development.

Duplications of the GH gene exist in all anthropoid primates (chapter 6), with placental expression documented for baboons, macaques, and spider monkeys. Humans are the only species for which data on the biological properties of placental GH exist. Placental GH has high somatogenic and low lactogenic activity, because GH2 has a low affinity (compared to pituitary GH1) for lactogenic receptors such as PRLR. However, the anthropoid placenta produces placental lactogens, which derive from descendants of a sequence of GH gene duplications and which do bind to PRLR.

In sheep, PGH affects placental and fetal physiology, but PGH production is largely restricted to early gestation (until day 50). Fetal pituitary GH expression begins around day 50 of gestation. In humans, GH2 acts on maternal and placental physiology but does not appear to affect fetal physiology directly, although if it regulates maternal IGF secretion, it would be involved in fetal growth via the IGF axis. Intrauterine growth restriction (IUGR) is associated with reduced placenta size, fewer placental cells expressing GH2, and lower maternal circulating GH2. In gestational diabetes mellitus (GDM), blood glucose is correlated with GH2 in maternal circulation. In a study comparing normal pregnancy with pregnancies complicated by either IUGR or diabetes, free GH2 in maternal serum, at both 28 and 36 weeks gestation, was correlated with birth weight (figure 7.1). Free GH2, IGF-1, and IGF-2 were all significantly lower in IUGR pregnancies at both time periods (McIntyre et al., 2000).

The GH genes of anthropoid primates have undergone accelerated evolution. For example, in the prosimian primates the lesser and thick-tailed bushbabies (*Galago senegalensis* and *Otolemur crassicadatus*, respectively), there is only a single GH locus, and bushbaby GH is highly conserved and differs little

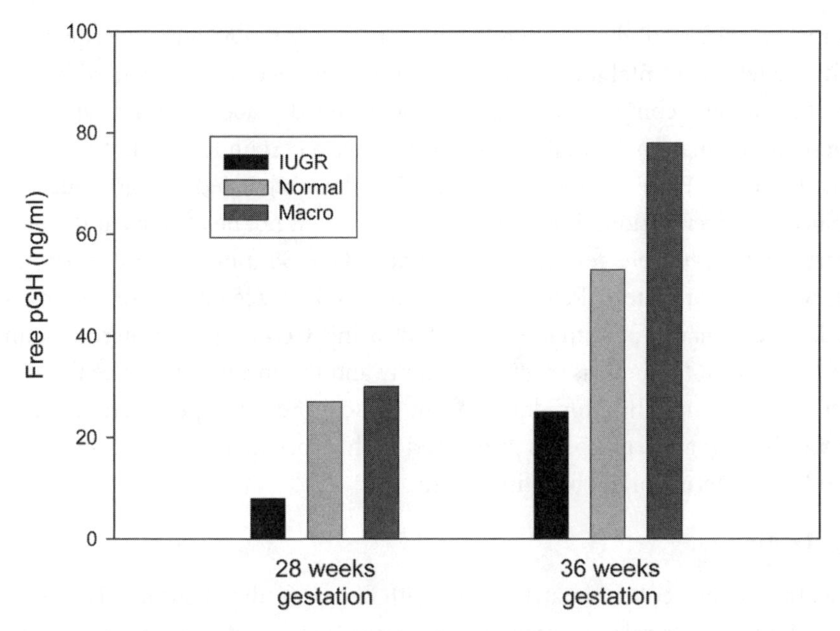

Figure 7.1. Free placental GH in maternal serum at gestational weeks 28 and 36, by growth category (IUGR < 10th percentile; Macro > 90th percentile). Data from McIntyre et al., 2000.

from that of nonprimate mammals. However, human and macaque GH differ from bushbaby GH by more than 60 residues (Adkins et al., 2001).

The evolution of the placental growth hormone system in primates and the assumption of an important aspect of regulatory control of maternal metabolism by the placenta have been suggested to represent an example of the maternal-fetal conflict hypothesis for placental evolution (Haig, 2008). Placental signaling via GH2 (and other molecules that induce lipolysis and insulin resistance, such as leptin, hCG, and CRH) serves to increase maternal circulating glucose, which benefits the fetus by increasing its access to glucose. Maternal glucose levels during normal pregnancy are positively associated with birth weight, explaining about 10% of the variation (Breschi et al., 1993). However, this increased circulating glucose may come at a cost to the mother under some circumstances. The downregulation of maternal pituitary GH, as well as maternal production of a binding protein for GH2, are hypothesized to represent maternal regulatory attempts to restrain circulating glucose (Haig, 2008). Of course, maternal downregulation of maternal pituitary GH secretion could also be viewed as an adaptation for regulatory efficiency. Variation in maternal blood glucose concentration probably has greater consequences for the fetal/placental unit than it does for the mother. As is evidenced by the birth outcome complications due to uncontrolled diabetes, either type 1 or gestational, maternal circulating glucose above a certain level is not adaptive for the fetus. Placental secretion of GH2 is downregulated at high maternal glucose concentrations, but that mechanism might not be effective if the maternal pituitary were continuing to produce GH. Shifting significant regulatory control of maternal circulating glucose to the placenta may represent an adaptive scenario for both mother and offspring. Certainly the maternal brain benefits from the increased glucose supply and the insulin resistance that reduces the competition for glucose from muscle and other peripheral organs. During pregnancy, glucose is prioritized to the placenta (and hence the fetus) and the maternal brain, benefiting both.

Leptin

Leptin is a molecule intimately linked with fat and feeding behavior. However, in addition to its role as a regulator of energy intake and adiposity, leptin appears to have important regulatory actions in many reproductive processes (Castracane and Henson, 2002). These functions include an association with the onset of puberty and roles in fertility for males and females, in ovarian folliculogenesis, and in implantation of the fertilized ovum. Leptin also appears to be involved with fetal growth and developmental processes. This illustrates another biological truism—biologically active molecules often have multiple functions and are employed in many physiological systems.

Because leptin is strongly associated with a measure of maternal nutritional status (fat mass), it is a plausible candidate for being an important metabolic signal for the maintenance and duration of pregnancy. Low leptin levels are associated with pregnancy loss in humans. Leptin levels may be abnormally high in pregnancies complicated by conditions such as diabetes mellitus and preeclampsia. Leptin is considered to be permissive for pregnancy, but not required. It may serve as a signal that the maternal condition is satisfactory for reproduction. Doubtless, since leptin is also produced by the placenta, abnormal leptin levels may also reflect abnormal placental function. For example, low leptin levels in pregnancy loss may reflect placental failure.

Leptin is produced by the placenta in many mammalian species, including humans, baboons, bats, rodents, pigs, and sheep. Significant differences in leptin regulation and function during pregnancy exist between rodents and primates. In rodents, the placenta largely secretes leptin into the fetal compartment and only minimally into the maternal compartment. In humans (and baboons) leptin is produced on both sides of the placenta; that is, placental production contributes to both maternal and fetal leptin concentrations (Henson and Castracane, 2002). Although most placental secretion is into maternal circulation, given the difference in size between the mother and her fetus, placental leptin contributes significantly to both. In humans, maternal serum leptin concentration is highest during mid-gestation and then declines. Pregnancy is considered to be a state of hyperleptinaemia with leptin resistance; i.e., high maternal leptin does not decrease food intake. Maternal leptin levels drop precipitously at parturition.

Leptin is associated with placental and fetal growth. Placental weight correlates with placental leptin mRNA; cord serum leptin is correlated with placental leptin mRNA, maternal serum leptin, and fetal mass (Jakimiuk et al., 2003). Fetuses that are large for gestational age have higher than normal leptin levels, and fetuses that are small for gestational age have lower leptin levels. In twin pregnancies, the larger twin has higher amounts of circulating leptin. In humans, cord blood leptin is associated with both length and head circumference of neonates. Leptin is also associated with insulin, insulin-like growth factor, and growth hormone, but it appears to be an independent predictor of fetal size in humans. In addition, leptin is found in amniotic fluid, and higher levels are associated with eventual preeclampsia; the highest levels were found in women with both preeclampsia and IUGR (Wang et al., 2010).

Evidence supports the hypothesis that most fetal leptin is of placental origin, although some is produced by fetal adipose tissue, especially at the end of gestation. Leptin receptors are found in the placenta. Interestingly, placental expression of leptin receptor is reduced in obese women (Farley et al., 2010). Human data are lacking, but in rodents, leptin receptors are found in many, if not most, fetal tissues (e.g., besides adopicytes, they are also noted in hair follicles, cartilage, bone, lungs, pancreatic islet cells, kidneys, testes, and so forth). Leptin receptors are found in fetal baboon lungs, and they increase in density at the end of gestation (Henson et al., 2004). Leptin may be a signal/marker of growth and development. In rats, leptin has significant developmental effects in the hypothalamus about one week after birth; developmentally, this would correspond to a late-gestation primate. Leptin receptors are also noted in the intestines of neonates, and leptin has been measured in milk, more evidence that early in life, leptin has developmental functions. Of note, in humans leptin levels in amniotic fluid are higher if the fetus is female (Chan et al., 2006), implying that some of the sex differences in fat metabolism (Power and Schulkin, 2008) may be set in utero.

Leptin may play a role in the fine-tuning of the timing of parturition in sheep. Intracerebroventricular infusion of leptin into late-gestation sheep fetuses inhibits the rise in fetal circulating ACTH and cortisol (Howe et al., 2002). Whether this effect is mediated through CRH pathways in the fetal brain, which also are involved in sheep parturition, is unknown. Restricting food intake below normal levels during pregnancy in sheep and rats results in increased adipose tissue, higher circulating leptin concentrations, and later, higher food intake in the offspring (Vickers et al., 2000).

Chorionic Gonadotropin

Chorionic gonadotropin is a member of a glycoprotein hormone family whose members (TSH, FSH, LH, and CG) have a structure composed of a common α unit and a unique β unit. As discussed in chapter 6, the β unit for CG is derived from a gene duplication of the β unit gene for LH from a common ancestor of all anthropoid primates. Levels of CG in human pregnancy are detectable as early as two weeks of gestation, peak near the end of the first trimester (around twelve weeks), and then decline to extremely low amounts (Jaffe et al., 1969). In accordance with this pattern, the known functions of CG are related to initiating and maintaining implantation. Although CG is produced by all anthropoid primates, most is known about the functions and variations in human CG. Accordingly, the text below is predominantly about CG during pregnancy in humans.

In early gestation, two forms of cytotrophoblast cells, villous and extravillous, are important for establishing the maternal-placental connection. Villous cytotrophoblasts cover the chorionic villi; they aggregate and fuse to form the multinucleated syncytiotrophoblast. As reviewed in chapter 3, the syncytiotrophoblast is the major endocrine tissue of the human placenta as well as the surface through which molecules must pass from the mother to her fetus. Extravillous cytotrophoblasts are derived from trophoblast cells at the tips of the anchoring villi. They proliferate and form columns of cells, which then invade the uterine myometrium. Some differentiate into multinucleated giant cells. Extravillous cytotrophoblast cells anchor the chorionic villi to the decidua. They also line the spiral arteries and are essential in the subsequent remodeling of these arteries, decreasing their elasticity to enable large amounts of slow-flowing maternal blood to enter the extravillous space. Both villous and extravillous cytotrophoblasts express CG, though its function differs between the two cell types.

During very early gestation of successful pregnancies, the form of CG predominantly produced by the human placenta is the hyperglycosylated type (Cole, 2007). This form of CG assists with maternal immune tolerance toward the extravillous trophoblast cells that invade the decidua to reach the maternal spiral arteries. It also appears to be involved in the tissue remodeling to transform the spiral arteries. Early pregnancy loss is associated with low maternal circulating CG, but even with appropriate total CG levels, proportionally lower circulating maternal serum hyperglycosylated CG (below 50%) is a risk factor for early pregnancy loss (Sasaki et al., 2008). The less-glycosylated form of CG binds to the receptor on the corpus luteum, providing the signaling necessary to continue the production of progesterone. This is critical in early-to-mid gestation, until the placenta begins producing sufficient progesterone to maintain the uterus in an appropriately quiescent and receptive state.

Because of its similarity in structure to TSH, at high concentrations hCG can bind to and activate the TSH receptor; this will depress TSH secretion but may result in excess thyroxine (T4) secretion. This causes a transient thyrotoxicosis, which resolves with the decline of circulating hCG after 12 weeks gestation and may be causally linked to morning sickness, at least in some cases (see chapter 8 for further discussion).

The evolution and function of hCG is an important area for understanding the evolution of human placental biology and fetal development. Not only is CG unique to the anthropoid lineage, but also, the human lineage appears to display the greatest number of functional CG genes from the various DNA duplication events that occurred throughout our evolutionary history. We discuss hCG in more detail in chapter 8.

Corticotropin-Releasing Hormone

Corticotropin-releasing hormone was first isolated from sheep hypothalamic tissue (Vale et al., 1981) but since has been found to be widespread among vertebrates. CRH is a member of a peptide family with at least four ligands (CRH, urocortin, urocortin II, and urocortin III), a binding protein (CRH-BP), and at least two receptors (three in catfish—see figure 7.2), each with multiple splice variants. Perhaps the most studied aspect of CRH function is its role as a neuropeptide. In both mammalian and nonmammalian vertebrates,

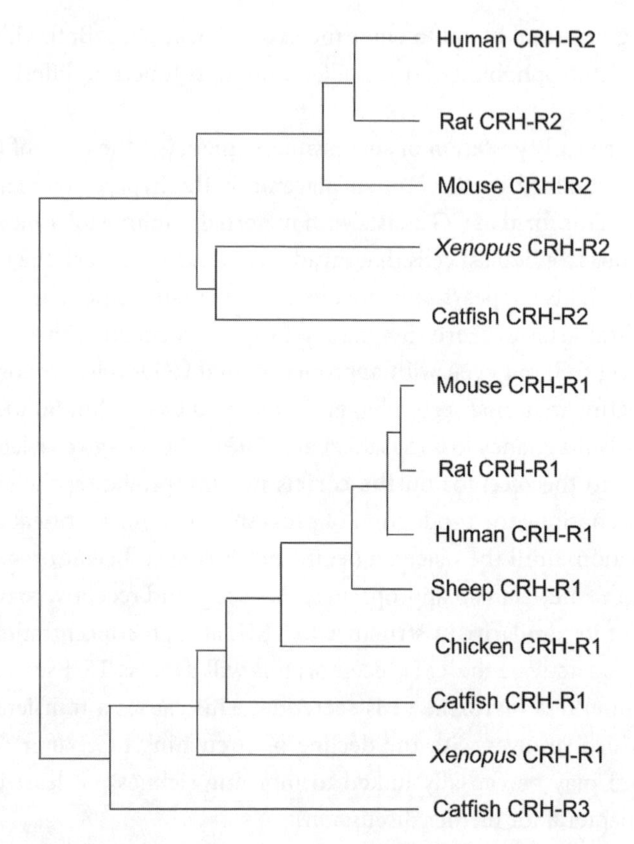

Figure 7.2. Phylogentic relationships for CRH receptors types 1, 2, and 3.

central nervous system CRH plays multiple roles in regulating and coordinating responses to external and internal challenges to viability, both through its effects on the pituitary and as a neurotransmitter/neuromodulator (Seasholtz et al., 2002). Released by the paraventricular nucleus (PVN) of the hypothalamus, CRH induces adrenocorticotropic hormone (ACTH) release from the anterior pituitary gland. In response to ACTH, the adrenal glands produce glucocorticoids (cortisol in many species, including primates; corticosterone in rodents and some other species), as well as other steroids. Release of CRH by the PVN is often associated with challenges to viability, such as detection of a predator or exposure to noxious stimuli. Central nervous system CRH also has direct and indirect effects on appetite, energy metabolism, blood pressure, arousal, and sexual behavior, among many other aspects of physiology (Dallman et al., 1995; Schulkin, 1999). The uses of CRH in behavioral organization are complex and diverse and not likely to be able to be placed under any single overarching cause, such as stress, though stressful phenomena will likely result in increased central nervous system expression of CRH.

The CRH Peptide Family. The CRH peptide family has been incorporated into many signaling functions throughout the body (Power and Schulkin, 2006). Urocortin and CRH have an affinity for CRH-R1 of approximately the same magnitude. However, the three urocortin ligands have a higher affinity for CRH-R2 than does CRH by an order of magnitude (Lovejoy and Balment, 1999). Urocortin II and urocortin III are thought to be selective ligands for CRH-R2, with low affinity for CRH-R1 (Lewis et al., 2001).

The CRH hormone family is ancient. The urocortins are related to urotensin-I, found in fish, and to sauvagine, found in amphibians (Seasholtz et al., 2002). The CRH binding protein is expressed in honeybees and is remarkably conserved. The ligands and receptors appear to share considerable homology with insect diuretic hormone; thus the origin of CRH appears to predate the split between invertebrates and vertebrates (Huising and Flik, 2005). A urotensin-I-like peptide has been found in the nematode *C. elegans*, suggesting that this peptide hormone family originated over 500 million years ago (Lovejoy and Balment, 1999). Human, rodent, carnivore and equid CRH have the same amino acid sequence, and the identity between mammalian and fish CRH ranges from 75% in tilapia and salmon to 95% in the suckerfish (Lovejoy and Balment, 1999; Chang and Hsu, 2004). Residues 9 through 21 are conserved among all CRH forms (Lovejoy and Balment, 1999).

CRH is not solely a neuropeptide; it is widely distributed in peripheral tissue, such as skin, liver, lung, and kidney (e.g., Emanuel et al., 2000; Slominski and Wortsman, 2000; Dotzler et al., 2004). The function and regulation of CRH varies remarkably among tissues. In peripheral tissue, it generally acts in an autocrine or paracrine fashion and is often associated with inflammatory processes, where its actions can be both pro- and anti-inflammatory (Slominski and Wortsman, 2000; Ilias and Mastorakos, 2003). It also appears to regulate energy metabolism and energy substrates in the periphery; CRH receptors are expressed in adipose tissue (Seres et al., 2004), and peripherally infused CRH increases energy expenditure and fat oxidation in humans (Smith et al., 2001).

CRH as a Reproductive Hormone. CRH is involved in many aspects of female reproduction, particularly implantation. Implantation is an inflammatory process in many species, especially those in which trophoblasts erode the uterine epithelium. Inflammation is associated with increased CRH expression. Expression of CRH is widespread throughout the female reproductive tract, where it appears to have paracrine actions related to inflammatory components of reproduction such as ovulation, luteolysis, implantation, and early maternal tolerance (Kalantaridou et al., 2004a, b). CRH is expressed in most female reproductive tissues besides the placenta, including the ovaries and the uterus of both pregnant and nonpregnant women (Clifton et al., 1998; Muramatsu et al., 2001; Zoumakis et al., 2001), and it performs a number of key roles in inflammatory processes related to reproduction, such as ovulation (Mastorakos et al., 1994; Kalantaridou et al., 2004a), decidualization (Zoumakis et al., 2000), and implantation (Athanassakis et al., 1999; Karteris et al., 2004). CRH receptors are expressed differentially in uterine tissue, depending on whether the female is pregnant or not (Grammatopoulos et al., 1998). Implantation sites in the uteri of rats in early pregnancy have high concentrations of CRH (Makrigiannakis et al., 1995); CRH is thought to act as an important paracrine signal in decidualization and implantation (Kalantaridou et al., 2010). In-vitro experiments on human trophoblasts indicate that CRH regulates the ability of trophoblasts to invade the decidua (Bamberger et al., 2006). Rats given a CRH-R1 antagonist during early gestation showed a dose-dependent reduction of endometrial implantation sites and live embryos (Makrigiannakis et al., 1995). However, interpretation of these results is complicated by the fact that CRH knockout mice produce normal-size litters (Muglia, 2000). This implies that the receptor (CRH-R1), not the particular ligand, may be the essential factor.

CRH in the Placenta. Shortly after CRH was isolated and characterized from extracts of sheep hypothalamus (Vale et al., 1981), it was detected in the serum of pregnant women (Sasaki et al., 1984). The CRH gene was subsequently shown to be expressed by human placental tissue (Grino et al., 1987; Frim et al., 1988) and to be the source of maternal (and fetal) serum CRH. Placental expression of CRH appears to be an anthropoid primate adaptation, as no other species are known to express CRH from their placenta (Power and Schulkin, 2006). Maternal circulating CRH has been detected during pregnancy in human beings (e.g., Goland et al., 1986), chimpanzees, gorillas (Smith et al., 1999), baboons (Goland et al., 1992; Smith et al., 1993), rhesus macaques (Wu et al., 1995), common marmoset monkeys (Power et al., 2006), and squirrel and owl monkeys (Power et al., 2010) but has not been detected in Madagascar prosimian primates (lemuroids) or any nonprimate mammals (Power and

Table 7.1 Presence versus absence of circulating CRH and CRH-BP during pregnancy in anthropoid primates and other mammals

Species	Circulating CRH	Circulating CRH-BP
Human	Yes; exponential increase	Yes; pattern known
Chimpanzee	Yes; exponential increase	Yes; pattern known
Gorilla	Yes; exponential increase	Yes; pattern known
Gibbon	Unknown	Yes; pattern unknown
Baboon	Yes; early-to-mid-gestation peak	Unknown
Rhesus macaque	Yes; pattern not known	Yes; pattern unknown
Common marmoset	Yes; early-to-mid-gestation peak	Yes; pattern unknown
Squirrel monkey	Yes; early-to-mid-gestation peak	Yes; pattern unknown
Owl monkey	Yes; early-to-mid-gestation peak	Unknown
Ring-tailed lemur	Not detected	Not detected
Brown lemur	Not detected	Not detected
Ruffed lemur	Not detected	Not detected
Rat	Not detected	Not detected
Guinea pig	Not detected	Unknown
Horse	Trace levels in pregnant and nonpregnant mares	Not detected
Sheep	Not detected	Not detected

Schulkin, 2006). There are no data on maternal circulating CRH during pregnancy in African and Asian prosimians (lorisoids and tarsiers), so the primate story is not yet complete; however, New World monkeys, Old World monkeys, and apes have all been shown to express CRH during pregnancy, and so far in no species of any kind outside of anthropoid primates has maternal circulating CRH been detected (table 7.1).

CRH in Anthropoid Primate Placentas. Because it acts in an endocrine fashion, placental CRH has the capacity to exert profound developmental and metabolic effects on both maternal and fetal physiology. Hypercortisolemia is characteristic of human pregnancy, and evidence supports the hypothesis that placental CRH is partly, though not solely, a regulator of maternal cortisol levels (Challis et al., 2000; Florio et al., 2002; Petraglia et al., 2005). Placental CRH, acting through its stimulation of cortisol from the maternal adrenals and also potentially via direct effects through the type 1 and type 2 CRH receptors, contributes to increased maternal metabolic rate and circulating glucose and to decreased maternal insulin sensitivity during pregnancy (Damjanovic et al., 2009). The existence of CRH receptors in many organs, especially the liver, kidneys, and GI tract, suggests that placental CRH may have multiple effects on maternal physiology and metabolism during pregnancy.

Placental CRH interacts with maternal physiology. For example, pregnant women who habitually go without food for 13 hours or more have higher serum concentrations of CRH than other women (Herrmann et al., 2001). A woman's tendency to engage in risk-taking behaviors is associated with higher CRH during pregnancy in both preterm and term pregnancies (Erickson et al., 2001). Elevated maternal circulating CRH during pregnancy is also associated with later changes in offspring physiology, such as an increased risk of central adiposity (Gillman et al., 2006) and, somewhat counterintuitively, higher levels of circulating adiponectin (Fasting et al, 2009). Adiponectin is linked with increased insulin sensitivity, while elevated CRH during pregnancy is associated with IUGR, higher cortisol exposure for the fetus, and an associated higher risk of eventual decline in insulin sensitivity and increased risk of metabolic syndrome. A speculative hypothesis is that the increased adiponectin in young children from high-CRH pregnancies represents an adaptive response to lower insulin sensitivity (Fasting et al., 2009).

Elevated maternal serum CRH is associated with adverse pregnancy outcomes such as preterm labor (Warren et al., 1992; McLean et al., 1995; Korebrits et al., 1998), fetal growth restriction (Goland et al., 1993), preeclampsia (Laatikainen et al., 1991; Perkins et al., 1993; Goland et al., 1995), and spontaneous abortion (Minas et al., 2007). Women who give birth prematurely exhibit both elevated CRH (Warren et al., 1992) and a precocious rise in CRH (McLean et al., 1995; Hobel et al., 1999; Leung et al., 2001). The mechanisms that link CRH with adverse pregnancy events are not well understood. Cortisol induces placental CRH production, so a link between stressful circumstances, preterm birth, and enhanced maternal circulating CRH is plausible. Progesterone injections have been shown to reduce the risk of preterm labor (Tita et al., 2010), at least among women at high risk (e.g., with a history of previous preterm birth.) In-vitro experiments have shown that CRH inhibits trophoblast production of progesterone, an effect that can be reversed by administration of a type-1 CRH receptor blocker (Yang et al., 2006). Thus an increase in placental CRH secretion might result in local progesterone withdrawal, leading to the initiation of labor. Progesterone competes with cortisol at the glucocorticiod receptor (Karalis et al., 1996), so exogenous progesterone theoretically would decrease placental CRH secretion and might block this effect. However, progesterone also interacts with a multitude of other receptors and pathways that potentially affect labor and parturition.

For most mammalian species, a key element in the progression of pregnancy to parturition is the conversion of progesterone to estrogen (Smith et al., 2005). As serum estrogen concentration rises during gestation, progesterone concentration usually declines. The anthropoid primate placenta does not express (at least to any appreciable extent) the enzyme 17α-hydroxylase-17,20-lyase (P450c17 or CYP17), which is required to convert progesterone to estrogen (Kallen, 2004). Thus, during human pregnancy, although maternal serum estrogen increases, progesterone does not decrease; progesterone is not the substrate for estrogen production in anthropoid primate pregnancy. Rather, estrogen is produced from the conversion of another steroid, dehydroepiandrosterone sulfate (DHEAS), produced by the fetal adrenal (Kallen, 2004; Rainey et al., 2004a, b).

These two aspects of human pregnancy, placental production of CRH and the existence of an extensive fetal adrenal zone that produces the necessary estrogen substrate (DHEAS) for normal progression of pregnancy, appear to be unique to anthropoid primates. These facets are linked, in that placental CRH stimulates the fetal adrenal to synthesize and release cortisol and DHEAS both directly (Smith et al., 1998; Sirianni et al., 2005) and indirectly, by stimulating the fetal pituitary to release ACTH (Lockwood, 2004). In pregnant chimpanzees and gorillas, maternal estradiol and CRH concentrations were highly correlated (Smith et al., 1999), consistent with the hypothesis that placental CRH drives placental estrogen synthesis through its stimulation of the fetal adrenal. Placental CRH is upregulated by cortisol (Robinson et al, 1988; Jones et al, 1989), forming a positive feedback system that, over the course of pregnancy, results in increasing placental production of estrogen via conversion of DHEAS (Kallen, 2004; Lockwood, 2004).

The maturation of the fetal HPA axis serves a vital function in the initiation of parturition for a number of species, including anthropoid primates (Lockwood, 2004; Smith et al., 2005). It seems that anthropoids possess an unusual, possibly unique, materno-feto-placental cooperative endocrine unit that functions to produce progesterone, estrogen, cortisol, and other hormones necessary for the normal progression of gestation (Kallen, 2004). For much of gestation, the fetal adrenal apparently does not express 3β-hydroxysteroid dehydrogenase / Δ5-Δ4 isomerase (HSD3β2), which means it cannot *de novo* produce glucocorticoids or mineralocorticoids. Fetal adrenal steroidogenesis is shifted towards DHEAS production in the fetal adrenal zone (Kallen, 2004; Rainey et al., 2004a). Placental progesterone is converted by the fetal adrenal to aldosterone (in the definitive zone) and cortisol (in the transitional zone)

(Kallen, 2004). The placenta converts fetal adrenal DHEAS (and, to a lesser extent, maternal adrenal DHEAS) to estrone, estradiol, and estriol. The maternal adrenal glands increase production of cortisol and aldosterone; this hypercortisolemia and hyperaldosteronism is counterbalanced by high serum progesterone, which competes for the glucocorticoid and mineralocorticoid receptors (Rainey et al., 2004a). The hypercortisolemia of pregnancy results in only a modest suppression of ACTH (Lockwood, 2004), probably due to stimulation of the maternal pituitary by placental CRH (Ruth et al., 1993; Bowman et al., 2001). In the fetal circulation, CRH, ACTH, and cortisol all increase over gestation (Lockwood, 2004).

Placental CRH is thought to affect the rapid growth and development of the fetal adrenal zone in the second and third trimesters, most likely through its stimulation of the pituitary to produce ACTH (Rainey et al., 2004a, b). Anencephalic fetuses do not develop a fetal adrenal zone, and these pregnancies are characterized by low maternal serum estrogen (Rainey et al., 2004b). At term, the fetal adrenal zone accounts for half of the mass of the fetal adrenal, which in turn is approximately equal in mass to a fetal kidney. After delivery, with the sundering of the fetal-placental axis, the fetal adrenal zone rapidly regresses, implying that a placental factor, or perhaps more than one, is necessary both for initiating and maintaining the rapid growth of the fetal adrenal zone late in pregnancy (Rainey et al., 2004a, b). There is no adult counterpart to the fetal adrenal zone, although anthropoid primates remain unusual among mammals in continuing to produce DHEA and DHEAS in the zona reticulus of the adrenal gland throughout life (Conley et al., 2004).

Patterns of Placental CRH Expression. There are at least two patterns of maternal circulating CRH during pregnancy: an ape pattern (humans, chimpanzees, and gorillas) and a monkey pattern (baboons, common marmosets, squirrel monkeys, and owl monkeys.) The ape pattern features an exponential increase in CRH starting in early-to-mid gestation (figure 7.3). The monkey pattern rise in CRH also begins in early-to-mid gestation, but maternal circulating CRH peaks during mid-gestation and then declines or plateaus (figure 7.4). The data for baboons (Goland et al., 1992), common marmosets (Power et al., 2006), and squirrel monkeys (Power et al., 2010) are consistent with the possibility of a final rise in maternal circulating CRH in the week preceding parturition; however, the pattern at the end of gestation has not been sufficiently resolved in any monkey species to determine whether this is the case.

The function of the early peak in maternal circulating CRH common to

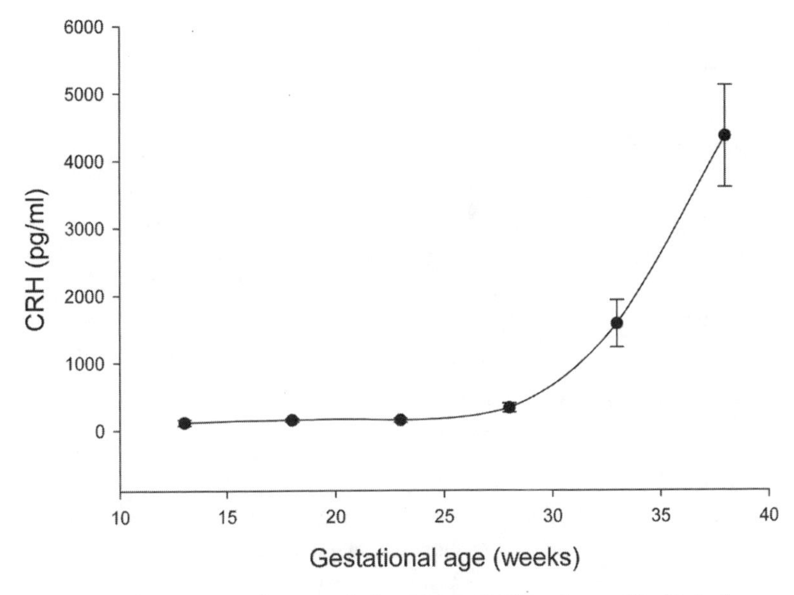

Figure 7.3. The pattern of maternal circulating CRH in the gorilla. Data from Smith et al., 1999.

all monkeys so far studied is unknown. In the common marmoset (*Callithrix jacchus*), peak maternal circulating CRH coincides with a period of significant fetal growth and development (Power et al., 2006), implying that placental CRH performs important functions beyond the endocrine circuit mentioned above that coordinates the progression and ending of pregnancy. These other possible functions may be lineage-specific (i.e., monkeys may differ from apes), but they also might reflect different functions for CRH at various times during gestation or for diverse target tissues. Placental CRH produced during gestation is likely to have multiple effects on many maternal and fetal organs besides the pituitary and adrenals, and these effects will probably be tissue-specific.

The consistency in the qualitative pattern of maternal circulating CRH found in an Old World monkey (the baboon; Goland et al., 1992; Smith et al., 1993) and now in three species of New World monkey (Power et al., 2006, 2010) suggests that it represents the ancestral pattern that arose before the last common ancestor of Old and New World monkeys. The pattern seen in great apes and humans thus is likely to be derived. Data from the smaller ape species (gibbons and siamangs) and from tarsiers and lorises would be helpful to further elucidate the evolutionary path of placental CRH expression in primates, as would data from more Old World monkey species, especially colobines.

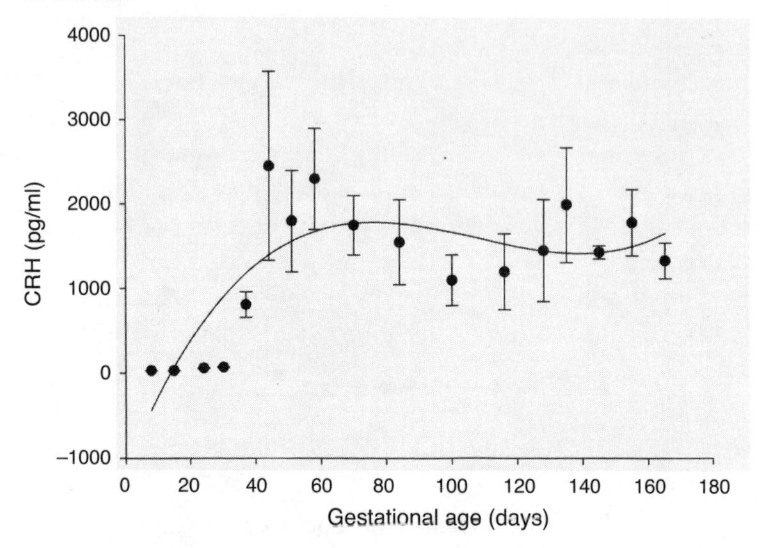

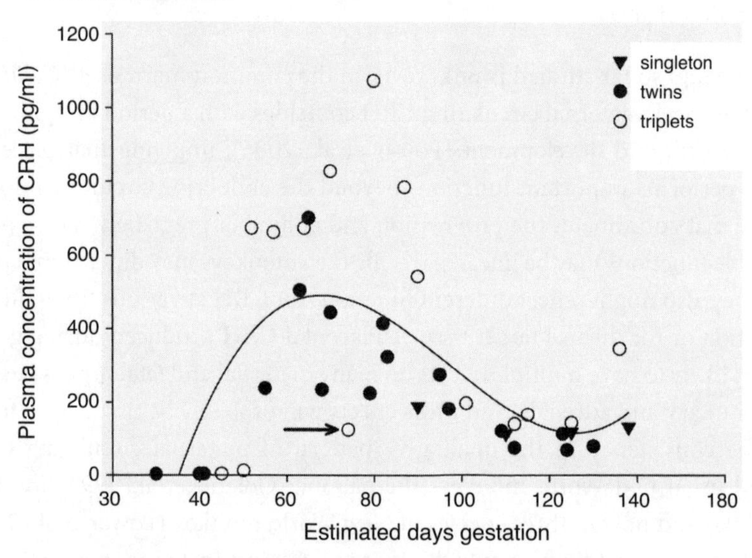

Figure 7.4. The patterns of maternal circulating CRH during pregnancy in (*A*) baboons (*Papio hamydryas*) and (*B*) common marmosets (*Callithrix jacchus*). The arrow indicates a value from a triplet pregnancy that subsequently aborted. Data in *A* from Goland et al., 1992; data in *B* from Power et al., 2006.

Serotonin

Serotonin (chemical name 5-hydroxytryptamine, abbreviated 5-HT) is a neurotransmitter with important functions in social behavior, including mating behaviors, but also behaviors related to feeding and sleep. Most serotonin is not found in the brain, however, but rather in the gut, where it regulates intestinal motility. Serotonin is also found in platelets, which take up excess gut 5-HT that is absorbed into the blood stream. Platelet 5-HT is active in blood clotting and acts as a vasoconstrictor.

Neural serotonin function is strongly associated with the processing of emotional information. A genetic polymorphism in the serotonin transporter gene is associated with biased attention toward emotional stimuli. People with the short allele exhibit biased vigilance toward negative stimuli, while people with the long allele show a bias toward avoiding negative stimuli and vigilance toward positive information (Fox et al., 2009). It is easy to conceive how both these tendencies could have adaptive value and how they would result in different temperaments in people.

Serotonin is important in fetal brain development (Bonnin and Levitt, 2011). Serotonergic neurons in the fetal hindbrain appear early in development and rapidly extend 5-HT axons into the forebrain. 5-HT signaling has been shown to influence cell migration and axon growth. Disruptions in 5-HT signaling in fetal mice are associated with long-term behavioral abnormalities, including increased anxiety in adult mice.

Serotonin is not a gene product; there is no 5-HT gene. Serotonin is produced by the metabolism of the amino acid tryptophan. Sources of 5-HT in the body are enterochromaffin cells in the gut and raphe nuclei in the brain. Raphe nuclei are present in fetal mouse hindbrain from early in gestation and appear to provide an endogenous source of 5-HT in that brain region. However, 5-HT is detected in fetal mouse forebrain days before raphe nuclei or any serotonergic axons reach those brain areas, implying the existence of a source of 5-HT outside the brain. The fetal gut and maternal blood were proposed as possible sources (Lauder et al., 1981, Yavarone et al., 1993); however, the fetal mouse gut does not begin to produce 5-HT until later in gestation, and less than 1% of 5-HT from maternal blood can be shown to cross the placenta (Bonnin et al., 2011). Using mouse knock-out models and in vitro tissue cultures, Bonnin and colleagues demonstrated that mouse (and human) placenta is capable of metabolizing tryptophan to 5-HT and that the placenta was the probable source of the 5-HT found in the forebrain early in gestation. In an elegant experiment, they demonstrated that when mouse placenta is presented with tryptophan on the maternal side, 5-HT is rapidly released into the fetal circulation (Bonnin et al., 2011, Bonnin and Levitt, 2011).

Postpartum Depression

We have spent a considerable amount of text emphasizing that the placenta is a critical regulatory organ that interacts with maternal physiology and metabolism. It is not surprising, then, to find that upon the loss of this regulatory organ at the birth of her child, a mother may feel not quite herself. Her physiology and metabolism had been undergoing considerable change throughout gestation and now must rapidly adjust again, this time to the lactating state. Depression is a leading cause of disability and affects women at two to three times the rate of men (Gjerdingen and Yawn, 2007). Postpartum depression (PPD) is a syndrome with a wide range of severity in symptoms that affects many mothers, with poverty being associated with higher prevalence (Segre et al., 2007). PPD, or the "baby blues," is thought to contribute to the higher rate of mood disorders among women (e.g., Hobfoll et al., 1995). This type of depression is not well understood. However, since it is related to pregnancy, it may have a placental element.

The loss of the placenta means the disappearance of a source of endocrine factors that were helping to regulate the mother's metabolism and were even a source of the signaling that started the cascades that prepared her mammary tissue for lactation. Post partum, she no longer has a placental source for growth hormone; her pituitary must restart producing and secreting GH after several months of being effectively shut down. She is no longer exposed to a peripheral source of CRH, which had been signaling her adrenal glands and pituitary for months. Progesterone and estrogen levels decrease. There are no more placental lactogens being released into her circulation. Her insulin sensitivity has changed. A mother's leptin levels also begin to decline, well before she loses any adipose tissue she gained during gestation.

One of the proposed advantages of placentophagy is to recover some of these information molecules no longer being secreted into maternal circulation. However, merely ingesting these molecules cannot replicate the complex, regulated physiological effects of their placental secretion into the maternal blood stream.

In-Utero Programming and Later Adult Disease

Fetal programming of physiology is a rapidly expanding area of biological knowledge that has critical relevance for human metabolic diseases. There are many potential mechanisms that can direct development of the fetus onto different paths, and these different developmental paths can have significant metabolic and physiological consequences many years later.

Many modern human diseases can be roughly classified as metabolic disorders and their sequelae. Obesity, diabetes, hypertension, cardiovascular disease, and so forth are quickly becoming the leading sources of mortality and morbidity across the globe. Epidemiological evidence, first from Forsdahl on differences in mortality rates due to heart disease among Norwegian counties (Forsdahl, 1977) and then by Barker in England (Barker, 1991), led to the hypothesis that early life experience, including in-utero experience, can lead to later predisposition to heart disease. This theory has been extended to other metabolic disorders, such as obesity and diabetes (Power and Schulkin, 2009).

One mechanism thought to underlie these relationships is fetal programming by nutritional stimuli. Fetuses may have to adapt to the supply of nutrients crossing the placenta, either a deficit or an overabundance, and these adaptations may permanently change their physiology and metabolism (De Boo and Harding, 2006). These programmed changes can have a significant impact on offspring later in life and may be the origins of a diverse array of diseases, including heart disease, hypertension, and noninsulin-dependent diabetes. In fact, because of fetal programming, obesity may become a self-perpetuating problem. Daughters of obese mothers may themselves be both vulnerable to becoming obese and more likely to have offspring that share this vulnerability, possibly a form of "inheritance of acquired characteristics" from mother to child. The placenta occupies a critical place in this transmission. Prior to birth, clues about the nutritional environment come from the mother, through the placenta.

The original hypotheses of Forsdahl and Barker were centered on poor nutrition and poor fetal growth as factors in predisposing adults to later metabolic disorders. The "thrifty phenotype" hypothesis posited that the initial metabolic changes in utero were a predictive adaptation, altering metabolism to better cope with an expected poor nutritional environment post partum. Pathology results when the ex-utero environment is not poor but rather provides sufficient or even excess calories—a mismatch between the predicted and the actual nutritional environment. Epidemiological evidence now shows that either end of the birth weight distribution entails increased risks of excessive weight gain, eventual obesity, and metabolic syndrome. Being large for gestational age has, ironically, some similar drawbacks to being small for gestational age. The proposed mechanisms for the programming of fetal physiology associated with maternal obesity include poor maternal glucose regulation and maternal dysregulation of cortisol metabolism. This condition is more one of pathology as opposed to an inappropriate adaptive strategy. At the least, the premature and substantial upregulation of insulin secretion by the fetal pancreas in response to poor maternal glucose control appears to lead, in many cases, to an eventual exhaustion of this organ later in life, an example of a phenomenon termed allostatic overload (McEwen, 1998, 2004).

Adipose Tissue as an Endocrine Organ

Fat is a good thing in our diets and on our bodies as long as it exists in moderation. Fat (lipid) is an essential part of our bodies and performs multiple and diverse functions, including providing nutritional, hormonal, and even structural support. The main fat depots in the body are in adipose tissue. Adipocytes are cells, specially adapted for fat storage, that provide energy for use at a later time and help to avoid the negative metabolic consequences of excess cellular lipid in organs such as muscle, liver, and heart. However, adipose tissue is not just a passive organ; it actively regulates metabolism through multiple pathways. A large number of cells other than fat cells are found in adipose tissue, including fibroblasts and immune cells such as mast cells, macrophages, and leukocytes (Fain, 2006). Adipose tissue synthesizes and secretes numerous peptide and steroid hormones as well as cytokines from both adipocytes and non-fat cells, and such factors are known to influence local and systemic physiology (Fain, 2006). In effect, adipose tissue forms an endocrine organ (Kershaw and Flier, 2004; Ahima, 2006).

Adipose tissue has endocrine function in at least three different ways. One, it stores and releases preformed steroid hormones. Two, many hormones are metabolically converted from precursors in adipose tissue, or active hormones may be converted to inactive metabolites. Adipose tissue expresses numerous enzymes involved in steroid hormone metabolism. For example, estrone is converted to estradiol in adipose tissue. Indeed, most if not all circulating estradiol in postmenopausal women comes from adipose tissue (Kershaw and

Flier, 2004). Adipose tissue produces both 11β-hydroxysteroid dehydrogenase type 1 (11β-HSD1), which converts cortisone to cortisol, and 5α-reductase enzymes, which convert cortisol to 5α-tetrahydrocortisol (5α-THF). Thus adipose tissue regulates local concentrations of glucocorticoids and contributes to metabolic clearance of glucocorticoids. Three, adipose tissue produces and secretes a large number of bioactive peptides and cytokines. These peptides are referred to collectively as adipokines to emphasize the role of adipose tissue in their synthesis and secretion.

It is the endocrine and metabolic functions of adipose tissue that cause much of the pathology associated with obesity. Just consider for a moment the metabolic and health consequences if a person's liver, thyroid, or adrenal glands doubled in size. Obesity is an increase in adipose tissue well beyond the tolerable functional range. In some aspects, the metabolic consequences of obesity are analogous to the endocrine dysfunction seen with hyperplasia or hypertrophy of an endocrine organ.

Since obesity and pregnancy both put unusual stress on the body, it is thus not surprising that obesity and pregnancy are not a healthy mix. Obesity during pregnancy is associated with greater mortality and morbidity of both the mother and her child (Leddy et al., 2008; Denison et al., 2010). Maternal obesity during pregnancy increases the risk of a number of complications throughout pregnancy and birth, including preeclampsia, gestational diabetes mellitus (GDM), and cesarean delivery (Leddy et al., 2008; Lynch et al, 2008; Denison et al., 2010). Maternal obesity is also linked with spontaneous abortion, both in spontaneous conceptions and those achieved through assisted reproductive technology. A meta-analysis of nine studies revealed that obese pregnant women have an estimated risk of stillbirth twice that of normal-weight pregnant women (Chu et al., 2007). Additionally, maternal obesity is a risk factor for a number of congenital defects, including spina bifida and other neural tube defects, hydrocephaly, cleft palate (Stothard et al., 2009), and heart defects (Mills et al., 2010).

Several mechanisms have been proposed for this relationship, including the increased risk of gestational diabetes and hypertensive disorders that are associated with obesity during pregnancy. Poor glucose control can contribute to congenital defects. Obesity and diabetes both result in multiple metabolic perturbations, such as dysregulated glucose and lipid metabolism. There may be multiple mechanisms by which obesity during pregnancy increases the risk of congenital defects.

Table 7.2 Expression of inflammation-related genes in human white adipose tissue and term placenta

Molecule	Adipocytes	Stroma Vascular Fraction of Adipose	Placenta
Adiponectin	++	0	0
Leptin	++	0	+
TNF-α	+	+++	+
IL-1b	NA	++	NA
IL-1Ra	+	++	+
IL-6	+	+	+
IL-8	+	++	+
IL-10	+	++	+
Resistin	0	++	+
MCP-1	+	++	+
MIF	++	+	+
PAI-1	+	++	+
VEGF	+	++	+
Cathepsin S	+	++	+

Modified from Hauguel–de Mouzon and Guerre-Millo, 2006.

The Placenta and Adipose Tissue

The placenta is also an endocrine organ intimately involved in regulating metabolism, signaling to both maternal and fetal tissue, and regulating metabolism, growth, and development of both. The placenta must interact and communicate with most, if not all, maternal organs. Adipose tissue is just one example, but given the effects of maternal obesity on fertility, pregnancy, and birth outcome, and the dramatic increase in maternal obesity over the last few decades (Kim et al, 2007), the interactions between the placenta and adipose tissue are a timely and instructive area to explore. In table 7.2, some of the molecules expressed by adipose and placental tissues are listed. These are gene products associated with inflammatory processes. Both obesity and pregnancy are associated with inflammation and inflammatory cytokines and other peptides.

It is therefore not surprising that obesity during pregnancy can lead to metabolic dysregulation. Both of these endocrine organs signal to maternal metabolism, but their signaling is not coordinated toward a common goal. Interestingly, there is a recent trend toward increasing placental weight in human pregnancy, which is in parallel with the increase in maternal weight (table 7.3; Swanson and Bewtra, 2008, Alwasel et al. 2011). Both of these potent endocrine organs are, on average, larger now than they have been in our past. It

Table 7.3 Placental weight increase in the United States

Year(s)	10th Percentile (g)	Mean (g)	90th Percentile (g)	Source
1959–1966	350	446	545	Naeye, 1987
1995	390	499	537	Swanson and Bewtra, 2008
2004	410	537	710	Swanson and Bewtra, 2008

would be fascinating to examine whether placental size is associated with any aspects of placental endocrine signaling. It certainly is reasonable to hypothesize that this increase in placental size is occurring at least in part due to the interactive signaling between the placenta and adipose tissue.

Inflammation

Pregnancy is an inflammatory state. Human pregnancy has been termed a state of low grade, albeit controlled, inflammation (Sacks et al., 1999). Pro-inflammatory cytokines are expressed by the uterine epithelium during the pregnant state. The underlying cause of the inflammation of pregnancy is not well understood. There are probably multiple effectors giving rise to increased monocyte activation during pregnancy. Possible triggers include a variety of molecules released into maternal tissue by the placenta (e.g., cytokines and anti-angiogenic factors) but also necrotic debris shed by the syncytiotrophoblast, including what are termed syncytiotrophoblast membrane microparticles (STBMs). These microparticles are continuously shed from the synctiotrophoblast into maternal circulation and have been shown to result in monocyte activation in vitro (Messerli et al., 2010). The inflammatory response results in an influx of leukocytes into uterine tissue. With normal pregnancy, the immune environment is skewed toward Th2 cytokine expression, especially in early pregnancy (Chaouat, 1999; Challis et al., 2009). A shift toward Th1 cytokine expression is associated with spontaneous abortion, preterm delivery, and preeclampsia (Challis et al., 2009). As normal pregnancy progresses, an overall decrease in pro-inflammatory cytokines and an increase in counter-regulatory cytokines occur (Denney et al., 2011).

Obesity is also a state of low-grade inflammation. Adipose tissue releases inflammatory cytokines, especially in the obese state. The combination of inflammatory signals from excess adipose tissue and from the placenta risks upsetting the regulatory balance. Obesity appears to increase the expression of pro-inflammatory cytokines from the placenta, just as it does from adipose tissue. Obese pregnant women exhibited higher gene expression in adipose tissue for interleukin-6, interleukin-8, and tumor necrosis factor-α (Basu et al., 2011). An increase in placental macrophages with pro-inflammatory cytokine expression has been demonstrated for obese pregnant women (Challier et al., 2008) and baboons (Farley et al., 2009). Inflammation, both from obesity and pregnancy, is linked to insulin resistance. Among obese pregnant women, circulating leptin, C-reactive protein, and interleukin-6 are increased. The pro-inflammatory signals have been shown to affect placental function by, for example, upregulating the system A amino acid transporter (Denison et al., 2010). As mentioned earlier, placental expression of leptin receptor is decreased in obese women (Farley et al., 2010), perhaps in response to the high maternal circulating leptin. Both obesity and pregnancy act to increase maternal circulation of glucose, free fatty acids, and amino acids. It is reasonable to hypothesize that the placenta is able to sense the maternal environment and react to it by modifying its own development and gene expression. A concern is that modern obesity is sufficiently removed from the normal state of our evolutionary past that these placental adaptations may not be effective and may even be counterproductive, when viewed in relation to the long-term health outcomes of the infant and the mother.

Gestational weight gain is another factor important to pregnancy outcome. Gestational weight gain is predominantly composed of fat, and that fat is preferentially deposited into central depots, both subcutaneous truncal adipose tissue and visceral fat (Bodnar et al., 2010). Visceral fat is most strongly associated with metabolic disorders and is an independent contributor to insulin resistance. However, prepregnancy obesity is more strongly associated with gestational diabetes than is excessive gestational weight gain (Bodnar et al., 2010).

Epigenetics and In-Utero Programming of Disease

Many of the changes to fetal metabolism that occur in response to the in-utero environment of maternal obesity are probably due to epigenetic changes, methylation, demethylation, or other changes to the structure of the chromosomal DNA, without any actual change of DNA sequence. These epigenetic changes can activate or silence genes by such mechanisms as recruiting methyl-CpG binding proteins, which then block transcription factor from binding to the promoter sites or change chromatin structure, enhancing heterochromatin formation (Jones and Takai, 2001; Heerwagen et al., 2010). This may be a fundamental mechanism for changing physiological set points, because it changes gene expression in tissue.

Regulation is the key to survival. Animals are constantly adjusting their physiology and metabolism in order to remain within the bounds of viability. As discussed above, a new development in genetics is that regulation also seems to be the key to viability at the level of DNA. Our updated understanding of genomics brings it closer to that of regulatory physiology. Many of the changes to fetal metabolism in response to the in-utero environment of maternal obesity are likely due to epigenetic changes, that is, changes in DNA expression via DNA methylation or demethylation, histone modifications, or other changes to the structure of the chromosomal DNA, without any change of DNA sequence. These epigenetic changes can activate or silence genes without changing the actual DNA sequence (Jones and Takai, 2001). The placenta is a hot spot for epigenetic regulation, being in general the tissue with the lowest overall levels of DNA methylation. The placenta also expresses a large number of imprinted genes, another form of epigenetic regulation.

Immunology of Pregnancy

In a placental mammal there is an intimate biological connection between the mother and the fetus. The extent of the connection varies by placental type, to be sure, but placental tissue, which is of fetal origin, is in contact with maternal tissue. This must have immunological consequences. The natural killer (NK) cells of the immune system arose approximately half a billion years ago. Thus, the ability to detect self from not-self arose hundreds of millions of years before any known adaptations for live birth. If an embryo is retained within the mother, then it must be protected in some way from the normal maternal immune response to foreign tissue. Somehow the maternal immune system must be deterred from rejecting the foreign entity represented by her fetus.

Clearly this problem has been solved by numerous species, our own included. However, many of the diseases of human birth have an immunologic axis. For example, preeclampsia and recurrent spontaneous abortion are both associated with an increased inflammatory response, a poor recruitment of particular NK cell types that are protective rather than destructive, and possibly an immune response to the foreign expressed genes from the father's genetic contribution to the fetus (see chapter 8).

Over a half a century ago, Peter Medawar, who received a Nobel Prize for his work on tissue transplantation in 1960, drew a logical comparison between the maternal-placental connection and an allograft (a transplant from a genetically-different member of the same species). This came soon after the discovery of the Major Histocompatibility Complex (MHC), which provided an important framework for understanding the mammalian immune system's ability to distinguish self from not-self and to reject not-self. The fetus (and hence the placenta) is not-self. Why isn't it rejected as would be the allograft? In actuality, the placental-fetal unit differs from an allograft in significant ways. The tissues of the placenta are not directly supplied by maternal blood vessels. There are layers of cells that separate these two genetically distinct organisms. However, the maternal immune system is exposed to fetal and placental cells. It is a thought-provoking question, even if the analogy is not perfect.

Of course, if the placenta and fetus were routinely rejected, then placental mammals would have never evolved; they would have been a failed evolutionary experiment. Therefore, we know there is at least one way to effect maternal tolerance of the placental-fetal unit. Possibly, as many ways exist as do extant lineages of placental mammals. In essence, this is a regulatory challenge. The placenta must function in some manner to regulate the maternal immunological response to become one favorable to implantation and subsequent placental growth and development. In the next sections we explore the evidence, largely from laboratory rodent and human studies, for the means by which the placenta is tolerated by the uterus.

Preparing the Uterus for Pregnancy

Prior to ovulation, the human uterine endometrium goes through a series of changes, required for successful reproduction. A blastocyst will not be able to successfully implant unless these alterations occur. The mucosal lining of the human uterus is transformed from endometrium to what is called decidua. This change is heavily influenced by the sex hormones, especially progesterone. This transformation is also associated with infiltration of the uterine endometrium by leukocytes. Thus, the immune system plays an important role in the decidualization of the uterine wall. Most of the infiltrating leukocytes are uterine natural killer (uNK) cells. Low levels of T cells are present, and B cells are largely absent. In mice, decidualization is dependent on uterine dendritic cells (Plaks et al., 2008), which have potent immune functions and are regulated by progesterone (Butts et al., 2010). Implantation is a state of inflammation, and even tissue necrosis occurs, at least with the highly invasive placental types, such as in primates. Implantation is a curious case of activation of some aspects of immune function and suppression of others.

Prior to implantation, the conceptus expresses signaling molecules that provoke an immune response from maternal epithelium. After implantation, the placenta expresses both signaling molecules and surface antigens, to which the maternal immune system reacts. The placenta and fetus express paternal antigens and also, quite possibly, fail to express maternal antigens, whose absence may cause an immune response, though that has not been demonstrated. Either way, a maternal immune response should be mounted in response to the presence of the fetal-placental unit, and certainly, a response is mounted. However, rather than a destructive response, it is a regulatory response. The immunological signaling machinery is co-opted to serve developmental functions and to facilitate maternal acceptance of the implanted conceptus.

It has been hypothesized that the maternal immune system is somehow down-regulated during pregnancy, at least within the uterus. However, the uterus is not immune-privileged; it contains a fully competent immunological response complement. The uterus can mount an immune response toward pathogenic microorganisms during pregnancy. Maternal T cells do acquire a transient tolerance to fetal antigens, specifically those that are paternally inherited. This has been well demonstrated in mouse models (Tafuri et al., 1995). Prior to pregnancy, these antigens were recognized and destroyed by the maternal immune system. It appears that during pregnancy, the maternal immunological "self" is modified. This is an active process, requiring information to pass between the placenta and the maternal immune system, and involves regulatory T cells. Notably, it has been shown that fetal hematopoietic stem-progenitor cells have a tendency to become regulatory T cells (Mold et al., 2010). In this way, the fetal immune system is biased toward tolerance.

Uterine Natural Killer Cells

The maternal immune system has to be involved in implantation and in immune tolerance of the semi-allelic placenta and fetus. The decidualization process is inflammatory as well as developmental. The placental extravillous trophoblast is the tissue that the human maternal immune system primarily encounters. This tissue deeply invades the uterine wall, and the invasion must be carefully regulated. Too deep or too shallow an invasion will result in different and harmful pathologies (see chapter 8). Trophoblast cells differ from somatic cells; for example, they express imprinted genes, endogenous retroviral products, and oncofetal proteins that are not expressed by tissues in the neonate. Old theories hypothesized that trophoblasts somehow made themselves invisible to the maternal immune system, and in some ways that is true. There is a completely different pattern of HLA (human leukocyte antigen) system expression in trophoblast cells compared to that in somatic cells. Trophoblast cells do not express the highly polymorphic HLA-A and HLA-B, which would interact with T cells. They also do not express HLA-DR, -DQ, and -DP molecules. They do express the less polymorphic HLA-C, as well as HLA-E, -F, and -G molecules, which interact with uterine natural killer (uNK) cells (Sargent et al., 2006). The maternal immune system response is activated by trophoblasts, and a uNK cell response is triggered, but the uNK cell type is protective rather than destructive. The trophoblast is not invisible to the maternal immune system; instead, extensive communication between trophoblast cells and the maternal immune system occurs. The maternal immune system is quite "aware" of the semi-allogenic conceptus in the uterus. Clearly, this communication is vital and is part of the mechanism that produces immune tolerance. An inadequate uNK response is associated with early pregnancy loss and with preeclampsia (see chapter 8). For example, decreased HLA-G RNA and protein are linked with preeclampsia (Goldman-Wohl and Yagel, 2009). The conceptus activates maternal immune tolerance by its signaling.

A large proportion of the cells found in the decidua during the first trimester are uNK cells (Goldman-Wohl and Yagel, 2009). Uterine NK cells have a particular expression phenotype different from peripheral blood NK cells, expressing CD56 but not CD16. They are present in pregnant decidua and nonpregnant uterine epithelium. In the first trimester of pregnancy, they are the most abundant decidual leukocyte (Lash et al., 2010). During early gestation (eight to ten weeks), uNK cells express many angiogenic growth factors, such as vascular endothelial growth factor (VEGF) and placental growth factor (PlGF), among others. By 12 to 14 weeks, secretion of these growth factors has decreased, with an increase in expression of certain cytokines, such as IL-1β and IL-6 (reviewed in Lash et al., 2010). Uterine NK cells do have cytotoxic activity, but it is downregulated during pregnancy. The syncytiotrophoblast has been shown to constitutively release exosomes (nanometer-sized, membrane-bound microvesicles), which contain signaling molecules that suppress uNK cell cytotoxicity (Mincheva-Nilsson and Baranov, 2010). Thus continual communication by many mechanisms, between the placenta and the decidua, regulates the maternal immune response.

Immunological Advantages of Pregnancy

Intriguingly, pregnancy has a beneficial effect on maternal rheumatoid arthritis, an autoimmune disease. Pregnant women with rheumatoid arthritis generally experience a progressive amelioration of the disease throughout gestation, but the disease returns within a few months post partum, with rare exceptions. Thus, some authors have proposed that the mechanism for creating tolerance to fetal antigens also underlies the regression of rheumatoid arthritis during pregnancy (e.g., Adams et al., 2007). Other autoimmune diseases, such as multiple sclerosis and Graves' disease, also generally show improvement during pregnancy. The mechanisms for these ameliorations are not well understood, but perhaps they are associated with regulatory T cells.

8

Modern Gestational Challenges

Modern human beings are an evolutionarily successful species. We have spread across the earth, and our population continues to increase at an impressive rate. It would seem that our reproductive adaptations are quite effective. Yet the processes of conception, implantation, gestation, and finally parturition in humans is not without its dangers and setbacks for both mother and fetus. Childbirth is still a leading risk factor for mortality and morbidity among women of childbearing age. Maternal mortality has recently slightly increased in the United States, to 17 deaths per 100,000 live births. In much of sub-Saharan Africa, the maternal mortality rate is in excess of 500 per 100,000 live births (Hogan et al., 2010).

A number of authors have commented that human gestation is fraught with several possible diseases and poor outcomes for both the mother and fetus that appear to be, if not unique to human beings, at least far more common in human pregnancy than in the gestation of other mammals. These conditions include: high rates of early pregnancy loss, nausea and vomiting of pregnancy, preeclampsia and related maternal hypertension syndromes, and preterm birth. All of these conditions involve the placenta to a greater or lesser extent. For example, the placenta is intimately involved in the etiology of preeclampsia, which is essentially a placental disease. Assuredly, many of these pathological events are related to placental development and function, and placental pathologies are associated with many poor birth outcomes. What evolutionary advantage might such a seemingly inefficient and pathology-ridden reproductive system have conveyed to our ancestors? Or are many of these reproductive problems incidental outcomes of adaptive changes to human reproduction that persisted despite the increased risks of ancillary disadvantages? The proximate mechanisms that underlie these placenta-related diseases and conditions are not well known. There has been much recent speculation and theorizing on possible ultimate explanations for these states. An important question to consider is to what extent these diseases are truly pathological and to what extent they reflect normal adaptive function that is being inappropriately, or possibly excessively, expressed.

In this chapter we examine human reproduction from an evolutionary perspective, focusing on pathology, especially as related to placental development and function; we also attempt to discern adaptive functions that may predispose us to vulnerability to pathology. For example, a vulnerability to preeclampsia would appear unlikely to have had adaptive function per se, but the adaptive changes necessary to support a deeply invasive placenta may have favored a developmental pattern that included an increased possibility for preeclampsia as a maladaptive consequence, outweighed by the selective advantages.

Fertility, Efficiency, and Child Rearing

Human female fertility is remarkably low, at least when calculated by the number of copulations often needed to produce a viable pregnancy. The mean time to pregnancy for sexually active couples not using contraception is more than six months. The number of ovulations required before pregnancy occurs appears to be intrinsically higher in women compared to our nonhuman primate kin. Estimates of the chances of a clinically recognized pregnancy occurring in any given cycle (assuming no contraception and a fertile male partner) are about 30% (Macklon et al., 2002). Thus, human reproduction appears to be inefficient. Of course, no one is suggesting that human reproduction overall is compromised by these conditions; human beings are the most fecund of hominoids, the phylogenetic unit that includes us and our nearest relatives, the great apes. We have populated the entire planet. However, our high level of reproductive failure begs for an evolutionary examination.

First of all, evolution does not always favor efficiency. Efficiency is a human construct; in many cultures (but not all, and not over all time) efficiency has been viewed as a desirable quality of human endeavors. In evolution, reproductive fitness is the currency of most import. Often, the more efficient reproductive strategy will provide the greatest fitness, but not always.

Being efficient at getting pregnant is not necessarily adaptive. Evolution rewards the genetic lineages with the greatest reproductive fitness, in other words, those that produce the most members of future generations. Pregnancy is certainly a necessary step in that process, but the goal of reproduction is not pregnancy. Rather, the goal is to produce offspring that survive to become sexually mature adults, capable of having young of their own. For primates, that is a long, involved process with many steps that the offspring must successfully accomplish. It does certainly all start with a fertilized ovum that successfully implants. The implanted blastocyst must then produce a competent placenta and undergo many stages of development, first as an embryo and then as a fetus. The fetus must also grow, a different if related process from development. If all goes well, a successful parturition will be cued by signals passing among the mother, the fetus, and the placenta, and a properly developed and well-grown infant will be born. If all the correct hormonal signals were given, the mother will produce milk, and the infant will begin its period of maternal nutritional dependence outside of the womb. In primates, lactation usually lasts longer than gestation, due in part to a prolonged weaning period, during which offspring actively learn from their mothers and from other group members.

In many mammals, weaning occurs abruptly. The most extreme example is the hooded seal, which gives birth to a large, precocial pup on unstable sea ice in the Arctic. The mother nurses her pup for just under four days, providing milk that is up to 60% fat; then she leaves. During the brief lactation, the pup has doubled its weight primarily by depositing almost all of the fat it obtained from its mother's milk (Oftedal et al., 1993). The pup will metabolize the nutrients obtained from its brief lactation, develop, and grow muscle mass, until it can swim. Then it will either begin to catch its own fish, or it will die of starvation.

This example definitely does not illustrate the usual primate way. A primate infant is typically dependent on its mother and often other conspecifics as well, for a long time. This is especially true for the ape lineage, from which we are descended. Primate infants continue to nurse long after they begin to eat solid food. Much of the solid food they obtain at first comes from older animals, either because the infants are allowed to take it, or in some species because other animals actively offer food to infants (Tardif, 1997). Even after the infants have become juveniles and are competent self-feeders, they may still be nursing; this is especially true for the great apes (orangutans, gorillas, and chimpanzees) and most probably for ancestral humans. In addition, access to food sites may depend on tolerance by older animals or the protection of the mother or other elders. It is a long time before a young primate is nutritionally self-sufficient, and longer still before it becomes reproductively active.

Consider reproduction from the viewpoint of one of our female human ancestors of many hundreds of thousands of years ago. If she gets pregnant, and the pregnancy is successful, then about nine months later she will produce an infant. She will nurse that infant for a considerable length of time, probably more than three years. She is unlikely to produce another infant until the first one is weaned. What the mean interbirth interval was for our ancestors hundreds of thousands of years ago is uncertain. For great apes, our closest relatives, interbirth intervals are five to eight years. Modern humans have been able to drastically shorten that time, due to technological and social mechanisms that enable our children to become nutritionally independent from milk and mothers at an earlier age; but in hunter-gatherer societies, it is still usually three or more years between successful births, if the prior infant survives. Interbirth intervals of modern women are unlikely to represent the mean interbirth times of our female ancestors. Through our technological progress, we have enabled birth intervals to become far shorter than would have been biologically possible in the past. A short interbirth interval in one of our ancestors was most likely related to a failed pregnancy or an infant death. The best estimate of mean birth interval for our distant ancestors is something between three and eight years, depending on how far back in time we are considering. The closer to the last common ancestor with the chimpanzees, the more likely the mean interbirth interval will be on the longer side of the estimate; the closer to modern humans, the more likely it will be on the shorter side. Of course, we cannot discount the possibility that the great apes have evolved to have longer interbirth intervals than were the norm for ape and human ancestors of seven million years ago. If that is the case, then the higher end estimate of mean interbirth period would decrease, but a reasonable estimate is still that interbirth intervals were on the order of three, four, or even five years.

The point of this exercise is to consider the idea that time is one of the most limiting resources to female reproduction for great apes, and, to a lesser extent, modern humans. A successful reproductive event will occupy a considerable length of a woman's reproductive life span, especially when we consider our long-ago ancestors, where the interbirth time was likely longer and our mean reproductive life span was probably considerably shorter. Given those circumstances, quality would be favored over quantity—not just the quality of the fetus/infant in a genetic sense, but the quality of the reproductive event as a whole. For human females, especially in the past, the chance of a successful reproductive event (rearing an infant to independence) was surely dependent on social factors in addition to the qualities of the infant and mother. From modern notions such as those expressed in *It Takes a Village* (Clinton, 1996) to arguments by evolutionary anthropologists (e.g., Hrdy, 2009), human child rearing has come to be seen as quite a cooperative endeavor. Reproductive success depends on assistance given to mothers and their infants and older offspring by a range of other group members. Grandmothers, aunts, uncles, fathers, and even unrelated members of the social group are important for enabling an infant to eventually grow up to be an independent adult. In modern society, we have decreased the immediate necessity of these other group members. Our children usually will not go hungry without their assistance. This may not have been the case many tens of thousands of years ago (Hrdy, 2009). Lactating females may have greatly benefited from social support that included some form of provisioning. Weaned, but still young and not fully independent children, would have benefited from help from adults in getting food. These social strategies to enhance reproductive success are suggested to have greatly influenced our cognitive abilities (Burkart et al., 2009). Stability of the social environment may have been vital for reproductive success. There may have been adaptive advantages for females to take their time about getting pregnant, at least in achieving a pregnancy that went to term.

Early Pregnancy Loss

Human reproduction is characterized by an exceptionally high rate of failure. Failure to implant and early fetal loss may affect as many as half of fertilized human ova. The mean length of time to pregnancy for a couple not using birth control is about seven months. The mean efficiency of human reproduction has been estimated at less than 30% (Zinaman et al., 1996). Even if fertilization occurs, many pregnancies do not succeed. There is a significant amount of early pregnancy loss, due either to failure to implant, inadequate implantation, or early developmental failure (Macklon et al., 2002). Early pregnancy loss occurs in 30% to as many as 50% of conceptions, generally before pregnancy can be clinically recognized. Even among clinically recognizable pregnancies, approximately 15% will not result in a live birth. The fate of a fertilized human ovum is far from secure (figure 8.1).

This reproductive inefficiency has puzzled many clinicians and biologists. Why should such an important aspect of reproductive fitness be characterized by such apparent wastefulness? Indeed, the clinicians' term for this early

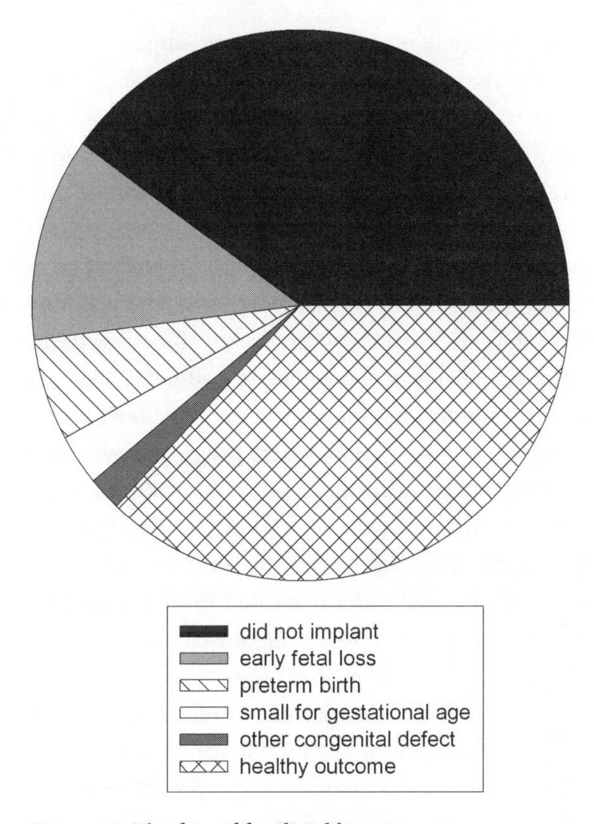

Figure 8.1. The fate of fertilized human ova.

pregnancy loss is fetal wastage, and there appears to be a large amount of fetal wastage in modern human reproduction (Simpson, 2007). This is usually presented as a medical issue and problem. Modern medicine is concerned with reducing this fetal loss where possible. Few researchers have investigated the possible evolutionary adaptive advantages of this so-called inefficiency and wastage. In other words, it is treated more as pathology rather than as an adaptive aspect of our past biology that we now wish to overcome.

A fertilized ovum needs to implant for proper growth and development to occur. Implantation is a complex process that requires extensive communication between maternal and fetal tissue. It also requires anticipatory changes in the uterine epithelium. The decidua is an altered state of uterine tissue necessary for implantation. Signals must come from the blastocyst, and certain maternal physiological changes are required to allow successful implantation.

In such a complex system, it is perhaps not surprising to find some failure. The question is whether there is significantly more such early fetal loss in humans compared to other species, especially our closest relatives. Undoubtedly, compared with many rodent species and domesticated animals, human reproduction appears fraught with lost opportunities. However, the more appropriate comparison is with our closer relatives, the Old World anthropoids. In two species of macaques, where early pregnancy was detected by monitoring serum CG concentrations, total pregnancy loss was about 17% over a ten-year period (Hendrie et al., 1996). The highest rates of loss were early in gestation or at the end of a fullterm pregnancy, due to stillbirth. Pregnancy loss within the first two or three months of gestation was at most ten percent. A 26-month study of chimpanzee birth outcomes, where pregnancy was determined by serum CG and confirmed by ultrasound in the first month of gestation, found that only 7 of 65 pregnancies (10.8%) failed during early gestation, with another 8 (12.3%) either resulting in mid-to-late-term abortions or stillbirths (Hobson et al., 1991). This admittedly limited evidence supports the hypothesis that human pregnancy has a high rate of early failure, compared to other closely related species.

What are the causes of this extensive early pregnancy loss in humans? Pregnancy loss can be categorized by gestational age, corresponding to major developmental stages. The earliest period is the pre-embryonic period, which lasts through five weeks after the last menstrual period. Pregnancy loss during this period probably represents implantation failure, and it is the pregnancy loss most likely to be associated with genetic defects (Warren and Silver, 2008). From six to nine weeks gestation is termed the embryonic period; after ten weeks and through the rest of pregnancy is called the fetal period. The rate of pregnancy loss is greatest in the pre-embryonic and embryonic periods.

A large proportion of fertilized human ova that fail to implant successfully and are spontaneously aborted have genetic mutations (Simpson, 2007; Warren and Silver, 2008). These aborted blastocysts/embryos often contain chromosomal abnormalities. The most common chromosomal abnormalities are trisomies, followed by monosomy X and polyploidy. Other chromosomal abnormalities that have been detected in early pregnancy loss at significant frequencies are chromosomal rearrangements, such as balanced reciprocal and Robertsonian translocations. Deletions of, and additions to, chromosomal DNA have also been detected. Finally, there are some known single-gene disorders that result in pregnancy loss.

Why should there be such a high rate of genetic abnormalities in human reproduction? This seems a puzzling phenomenon, and, of course, one rife with speculation. Was a high rate of chromosomal rearrangements adaptive in our past? Is that one of the factors that led to our evolutionary success? Did it provide genetic diversity that enabled our divergence from the ape clade? These are difficult questions to answer. Certainly the genetic differences between humans and chimpanzees identified by the human genome project don't appear to be inordinately associated with chromosomal rearrangements. However, rearrangements of significant segments of DNA have occurred. Chromosomal rearrangements are a potential source of variation upon which evolution can act, but in most cases they simply result in pregnancy failure.

The mechanisms by which these chromosomal abnormalities arise are unknown. They do not appear to be solely caused by defects in meiosis. Instead, it seems that aneuploidy can occur during mitosis in the early blastocyst, leading to a mosaic of normal and abnormal cells (Bielanska et al., 2002). An examination of 23 fertilized ova from IVF at the 8-cell stage determined that only 2 were completely normal (Vanneste et al., 2009). About half had no cells with a normal chromosomal arrangement. The rest were a mosaic, with some abnormal and some normal cells. Many blastocysts had multiple cells with different chromosomal abnormalities. Because these embryos were created by IVF, we should be careful in assuming that they are representative of in-vivo blastocyst development. These embryos were from healthy, young couples at risk for inherited genetic diseases unrelated to fertility, so this is consistent with the high proportion of early pregnancy loss due to chromosomal abnormalities.

A speculative hypothesis is that the increased early pregnancy loss in humans might serve as an adaptive mechanism to delay pregnancy. Ancestral females who suffered higher rates of early pregnancy loss may have experienced better long-term reproductive success. At first glance that may appear paradoxical, but the typical anthropoid reproductive strategy does not punish short-term reproductive failure to the extent that would occur in, for example, most rodent species. This is especially true for the ape female reproductive strategy, where a female will give birth to at best a few young over a long reproductive life span. Neither humans nor any of our great ape relatives display any seasonality in reproduction, so there have been no selective pressures for pregnancy to be efficient in order to meet some seasonal time constraint. Females can get pregnant at any time of the year. Whether a female gets pregnant in 1 cycle or after 5 or 6 may not be critical in species whose reproductive success depends upon the outcome of 6–7 births over a 20–30 year reproductive life span. The high rate of chromosomal rearrangements in human reproduction may represent an evolved mechanism for pregnancy failure, extending the number of cycles required, on average, for a successful pregnancy to occur. As we proposed above, efficiency in getting pregnant may not be the most adaptive strategy for producing a viable infant that will survive to independence.

Early pregnancy failure occurs due to a failure of implantation. Implantation is a complex, highly coordinated process. It requires signaling between the blastocyst and maternal tissue. Many axes in this process are vulnerable to disruption and could cause the process to abort. For example, production of progesterone by the ovary (the corpus luteum, specifically) is necessary to maintain the uterus in a receptive state to support implantation. The human placenta also produces progesterone, aiding its own cause; but perhaps most importantly, placental trophoblasts produce chorionic gonadotropin (hCG), which stimulates the corpus luteum and keeps ovarian progesterone production high. Thus, placental signals are required both to initiate and to sustain implantation. Progesterone and hCG are vital signaling molecules for pregnancy success, although they are not the only ones. The complex communication dance between blastocyst and decidua is susceptible to breakdowns at numerous points. Thus, it is not surprising that blasotocysts with chromosomal abnormalities usually fail to signal correctly and are aborted.

But, not all of these blastocysts abort; some chromosomal abnormalities appear not to compromise the appropriate (or at least adequate) signaling capacity of the blastocyst and placenta, and these fertilized ova do successfully implant and can result in term pregnancies. For example, trisomy 21 is one of the most common chromosomal abnormalities among live births, with a rate of about 1 per 1,000 live births. Trisomy 21 results in Down syndrome, although a small percentage of Down syndrome babies do not show trisomy 21. Maternal age is the leading risk factor for trisomy 21, with the incidence increasing rapidly with age, from 1 in 900 live births for women age 30, to 1 in 400 live births for women age 35, to 1 in 12 live births for women age 49.

The placentas of trisomy 21 pregnancies secrete hCG at a high level. This potentially produces a strong signal of a viable pregnancy to the maternal uterus and ovaries. There are two forms of hCG produced by the placenta, which differ in the amount and type of oligosaccharide residues attached to the molecule. In a trisomy 21 pregnancy, the hyperglycosylated form of hCG is preferentially produced. This form is far less biologically active in terms of stimulating progesterone from the corpus luteum; however, it does strongly enhance implantation (Cole, 2007, 2009). Thus, even though most chromosomal anomalies result in early pregnancy loss, trisomy 21 results in successful implantation. This is an example of a mutation that has selective advantage, due to its ability to enhance embryonic and fetal survival despite probable reduced survival and reproductive success of the eventual adult. Interestingly, trisomy 21 is also associated with a higher risk for extreme nausea and vomiting of pregnancy, which we discuss below.

Morning Sickness

Nausea and vomiting during pregnancy, or morning sickness, is a common experience for pregnant women, especially within the first trimester of pregnancy. Despite the colloquial name, the symptoms can and do occur at all hours of the day. For most women, it is an unpleasant but transient experience of pregnancy. Symptoms begin quite early, usually by five weeks after the last menstrual period, and they generally abate between the fourteenth and seventeenth weeks of pregnancy. Although undoubtedly unpleasant for the women in the short term, nausea and vomiting of pregnancy usually will have few long-term consequences for their pregnancy or their life. Morning sickness could be considered a normal part of pregnancy. Indeed, nausea and vomiting in the first trimester of pregnancy is associated with a decreased risk of many potential problems with pregnancy—miscarriage, preterm delivery, low birth weight, stillbirth, and fetal and infant mortality (Broussard and Richter, 1998). It is a sign of a viable pregnancy.

It isn't always transient, unfortunately. Some women experience symptoms throughout pregnancy. One study found that during the third trimester, 16% of women experienced nausea and 7% vomiting (Knudsen et al., 1995). The range of severity of symptoms is extremely broad. Some women suffer no nausea at all yet still deliver a healthy, term baby. Unfortunately, for a significant number of women, the nausea and vomiting can become seriously disruptive to their lives. Surveys have found that as many as a quarter of pregnant women suffer sufficiently severe symptoms to cause them to change their usual activities (e.g., O'Brien and Naber, 1992). The ancillary effects of severe nausea and vomiting of pregnancy can negatively impact a woman's work, her relationship with her partner, her other social relationships, and her daily life. A small percentage of women elect to terminate their pregnancy due to the myriad effects of severe nausea and vomiting (Mazzotta et al., 2001). For a few percent of women, the symptoms can be so extreme, with violent, sustained vomiting (hyperemesis gravidarum), that they may require hospitalization. Hyperemesis gravidarum is second only to preterm labor among reasons for hospitalization during pregnancy (Gazmararian et al., 2002).

Unfortunately, there are no reliable indicators that predict which women are likely to suffer from severe nausea and vomiting. Fatigue, multiple gestation, and molar pregnancies are risk factors (Broussard and Richter, 1998). Women who have previously had severe nausea during pregnancy, or whose mothers suffered from severe nausea during pregnancy, are at greater risk; it would appear that susceptibility to hyperemesis gravidarum has a heritable component. Women whose mothers experienced hyperemesis gravidarum are three times more likely to have a pregnancy complicated by severe morning sickness (Vikanes et al., 2010). A case control study found that women who experienced hyperemesis gravidarum were more likely to have had a sister and/or mother also affected with severe morning sickness. They also reported high levels of severe morning sickness in both maternal (18%) and paternal (23%) grandmothers (Zhang et al., 2011).

The cause of morning sickness is not known, though there are many theories. An old, now discredited theory postulated that morning sickness was one of the many hysterical female disorders. Nausea and vomiting of pregnancy was considered a conversion disorder—a hysterical expression of unconscious conflict, in which the women resented or were at best ambivalent about their pregnancies. Vomiting represented an oral attempt to abort the pregnancy. Other studies have examined links between severe morning sickness and psychiatric problems and levels of psychosocial stress. No associations with depression, bipolar disorder, schizophrenia, or anxiety disorders have been found. Not surprisingly, women with severe nausea and vomiting report higher levels of stress, anxiety, and depression; but that is thought to be more likely a result of the severe morning sickness than a cause of it (Verberg et al., 2005). Modern evidence favors the idea that morning sickness has an organic cause and is only marginally associated with psychological disorders, more as a contributing factor to psychological stress than as a consequence.

What does cause morning sickness? We don't really know, which is somewhat amazing given how ubiquitous it is, how many women it affects, and how many women undoubtedly wish modern medicine had a helpful answer. Suggested proximate causes of morning sickness include reactions to the alterations of maternal physiology or to the hormonal milieu of pregnancy. Because molar pregnancy (in which there is abnormal placental growth and little fetal tissue or even a complete absence of the fetus) can also result in morning sickness, often severe, the cause is likely due to a placental signal and does not come from the fetus.

Another puzzling question is, why should morning sickness exist? It seems somewhat strange that an early symptom of pregnancy is to make the woman feel miserable and possibly to reduce her food intake. What adaptive purpose might that serve? In many species, including some primates (e.g., common marmosets; Tardif et al., 2005), weight loss during early pregnancy is strongly linked to pregnancy loss. But, intriguingly, in humans morning sickness appears to be associated with a decreased risk of early miscarriage (Weigle and Weigle, 1989; Davis, 2004). Some researchers have even suggested that reducing maternal food intake early in pregnancy may have beneficial metabolic consequences for the early developing embryo, by enhancing placental growth (Huxley, 2000).

Several authors have risen to the challenge and proposed evolutionary adaptive theories to explain this puzzle—not the miserable part, but rather the change in appetite and feeding behavior in early pregnancy (Hook, 1976; Profet, 1992; Sherman and Flaxman, 2002). The hypothesis centers on the notion that the nausea of early pregnancy causes women to avoid eating certain types of foods during that crucial period. The theory is that certain foods contain substances, either chemical or microbial, that pose a risk to the developing fetus. For example, many plant foods contain secondary compounds that might be teratogenic, abortifacients, or otherwise harmful to the fetus. And inarguably, spoiled meat or fish might be contaminated by harmful bacteria. Pregnant women who were more likely to avoid these types of foods because of their queasy stomachs may have, on average, produced healthier offspring and remained healthier themselves. This has been termed the "maternal and embryonic protection hypothesis" for the adaptive function of morning sickness (Flaxman and Sherman, 2008).

How well does this evolutionary theory regarding morning sickness hold up? It is consistent with the timing of most nausea of pregnancy. Assuredly, during early pregnancy the fetus is most at risk from teratogenic substances. And since most morning sickness is mild (although women experiencing it may not think that is a proper description) and has little long-term effect, it is more properly thought of as a consequence of pregnancy rather than as sickness. The still-unresolved question is whether the consequences of nausea and vomiting of pregnancy had beneficial effects on female food choice, and if so, was that the selective pressure that brought it into our biology? This theory does not explain why nausea and vomiting of pregnancy should be so uniquely human. There is little evidence that other mammals are affected to anywhere near the extent that human females are, and yet many mammals ingest foods that contain potentially teratogenic substances or other materials that could be harmful to a fetus. Evidence suggests that the natural variation among human populations in the incidence and severity of nausea and vomiting of pregnancy is associated with dietary patterns, consistent with the food-avoidance hypothesis (Pepper and Roberts, 2006). In other words, populations that consumed more meat, sugar, alcohol, and what Peppers and Roberts (2006) termed "oil crops" and low amounts of cereals had higher rates of nausea and vomiting during pregnancy. However, the maternal and embryonic protection hypothesis appears weak in its ability to explain why the phenomenon arose in humans in the first place but apparently not in other species.

We now turn to a placental theory for a proximate cause of morning sickness. The placenta produces many bioactive molecules that affect maternal metabolism. Morning sickness is, in a significant sense, a metabolic disruption. One hormone in particular has drawn considerable attention regarding morning sickness: hCG. The symptoms of morning sickness are consistent with thyrotoxicosis. The structure of hCG makes it capable of activating the thyroid gland, due to a weak but significant ability of hCG to mimic the action of thyroid stimulating hormone (TSH) on the thyroid. During early pregnancy, hCG increases and TSH decreases, with hCG reaching its maximum at about the same time TSH is at its nadir (Glinoer et al., 1990). The evidence indicates that hCG is the primary ligand activating the thyroid gland in early pregnancy. Free thyroxine (T4) levels consistently increase in early pregnancy, in association with increased hCG but decreased TSH (Haddow et al., 2008). At high concentrations, hCG can bind to and activate the TSH receptor; this will depress TSH secretion, but it may result in excess T4 secretion. This causes a transient thyrotoxicosis, which resolves with the decline of circulating hCG after 12 weeks of gestation. Variability in hCG structure affects its ability to stimulate the TSH receptor; hyperglycosylated hCG is less effective (Cole, 2009). Baseline levels of TSH also appear to affect the ability of hCG to bind to the receptor, possibly by competition.

An important point relevant to evolutionary explanations of phenomena is that an evolutionary explanation does not imply that the phenomenon per se has or had adaptive value. Morning sickness may be a case in point. If the primary proximate cause of nausea and vomiting of pregnancy is the interaction between placentally produced hCG and the maternal thyroid gland, then the evolutionary explanation revolves around the adaptive function and selective pressures regarding hCG. Morning sickness may be merely a phenotypic outcome of selective pressure on the level of hCG production and its signaling strength.

The uterus requires progesterone to remain in a receptive state. Without sufficient progesterone, implantation will fail. Placentally produced hCG is the mechanism to stimulate progesterone production and secretion by the ovaries, so hCG production by the placenta is a necessity for successful pregnancy. It is a signal to the maternal system that a viable conceptus has implanted. Thus, it is a signal that likely is susceptible to different selective pressures with respect to maternal and fetal biology. For the fetus, the signal is absolutely vital—without it, the fetus dies. For the mother, a failed pregnancy is a possible loss of fitness, but not necessarily, because the loss would occur very early in pregnancy, before any significant maternal resources are committed. Thus, selective pressure to enhance placental hCG production would be strong, from the offspring perspective. For the mother, the strategy may be somewhat passive—prepare the uterine lining to receive a blastocyst, maintain that state for a period of time, then revert to the nonpregnant state if the appropriate signals are not received. For the blastocyst there is no tomorrow; either it implants or it dies.

We now have the makings of an evolutionary arms race, though it is possibly asymmetric, with a much stronger imperative on the part of the offspring. There is also a paternal interest that should be considered. Although the mother may not value any one blastocyst over another, paternity is less certain. Genetic techniques have shown that most so-called monogamous species display significant levels of cuckoldry (in both directions), and the extent of monogamy in our ancestors is quite uncertain, in any case. Each fertilized ovum may have a higher value to the male than to the female, as the next one may not be his. Therefore, paternal genes that enhanced placental signaling conducive to implantation would be favored. In essence, from the paternal and offspring perspective, evolution would favor genes that enhance placental signaling. Whether or not evolution would favor increased discrimination by the mother, either by requiring higher levels of signaling or more complex and coordinated signaling, is unclear, as the various costs and benefits to her fitness are complicated. Certainly the mother benefits from a competent blastocyst implanting, and in contrast would suffer a cost if an ultimately unviable fetus were carried for long. However, if female reproductive success were enhanced by delaying successful pregnancy for multiple cycles on average (due perhaps to the future social benefits accrued by establishing a longer-term association with one or more males), then her fitness strategy is in conflict with that of any individual blastocyst and not completely aligned with the male strategy, even if she does form a long-term association. Her benefit from any single competent blastocyst is lessened, while her cost due to carrying a nonviable fetus, or even one with low viability, is increased. This should create a scenario that would place great selective pressure on placental signaling and indeed initiate an "arms race" between maternal and fetal-paternal interests.

This scenario is consistent with the facts regarding human early pregnancy loss, morning sickness, and even menstruation and preeclampsia (discussed below). A conflict in strategies between the mother (who we hypothesize benefits from delaying pregnancy by being "inefficient"), the fetus (who has only this one chance), and the father (who cannot be certain future fetuses are his) has resulted in selection for strong signaling by blastocysts/embryos and a high incidence of defective embryos, whose signaling is impaired and therefore cannot successfully implant. Morning sickness may be a by-product of the selection for strong signaling due to one (at least) of the signaling molecules affecting maternal metabolism in such a way as to cause (unintended) nausea.

Of course, consistent does not mean correct. We are merely outlining an admittedly speculative hypothesis that incorporates multiple unusual aspects of human female reproduction and that explains morning sickness without attributing selective advantage to it per se. Once this scenario comes into being, and nausea during early pregnancy occurs as a by-product, selection might favor retaining it, if it truly did guide female feeding behavior into more adaptive patterns.

Why Do Women Menstruate?

This question has been asked for thousands of years, probably tens of thousands. We do know that Aristotle and Galen both weighed in on the matter. Aristotle conjectured that the menstrual blood was actually the substance that formed the fetus, under the guiding direction of the semen. Galen held that menstruation represented a flushing-out of bad humors from women. Indeed, menstruation was considered analogous to bleeding as a therapeutic tool to combat disease.

Modern science has a different conception of menstruation, though there have been suggestions that reflect the Galenic idea. For instance, menstruation serves to deplete the body of iron (Sullivan, 1981), and thus menstruation is one reason premenopausal women have lower cardiovascular disease risk than do men (high iron stores are a risk factor for heart attacks). The more generally accepted understanding of menstruation is that it is a consequence of the necessary changes that the uterine lining must undergo to become hospitable to a blastocyst (a hormone-driven modification called decidualization). If a decidua doesn't form, the blastocyst will be unable to implant, and the pregnancy fails. The decidua secretes many hormones and information molecules, and it has receptors for these hormones. The decidua is an important component of the machinery necessary for the maternal-placental-fetal crosstalk that maintains and directs gestational development. If there is no blastocyst, or the blastocyst fails to provide the appropriate signaling, then the hormonal and immunological signals that maintain the decidua cease, the modified tissue is sloughed, and the uterus returns to its nonreceptive state. Thus menstruation is not perceived as adaptive per se; instead, it is a necessary consequence of adaptive changes to the uterine lining that cannot be maintained indefinitely (Finn, 1998).

There have been other attempts at suggesting adaptive function to menstruation, hypotheses to suggest that it is more than just a consequence of the invasive implantation of anthropoid primates. Margie Profet (1993) questioned why natural selection would favor a process that depleted reproductive females of energy and nutrients. She suggested that selection should have minimized menstrual losses unless there was some adaptive function to a high blood flow. Her suggestion was that the female reproductive system needs to be protected against external pathogens that can be introduced via sexual intercourse and hitchhike on sperm. Menstruation could be a means of cleansing the uterus of pathogens, many of them sperm-borne, by flushing them out with the menstrual blood.

This hypothesis doesn't withstand careful scrutiny, but it does raise several important points. One, the female reproductive tract certainly does need defense mechanisms against pathogens. The uterus is one of the body organs that, as a necessity of its function, will be exposed to external pathogens and antigens, not as constantly as the lungs or the digestive tract, to be sure, but nonetheless, to perform its function the uterus must allow external matter inside. But some of that external matter (sperm, seminal fluid, and, most important, the blastocyst), while antigenic by definition, should be protected from immunological attack. The uterine immunological defenses must be able to distinguish pathogens from fitness-enhancing reproductive material. The early suggestion that the maternal immune system is suppressed in order to tolerate the fetus does not appear accurate, because there are robust maternal immune responses to pathogens. The tolerance of the fetus is not mediated primarily through T cells but rather through aspects of the innate immune system (uterine natural killer cells) and represents altered signaling, not immunosuppression.

The second important point is that the amount of blood lost in human menstruation, while not large (10–80 ml is considered normal), is significant. It is roughly 1% of the total blood volume, certainly in no way immediately threatening to health, but a curious waste of resources nonetheless. And while all anthropoid primates menstruate, the amount of blood lost is dramatically less in many species. For example, in vervet monkeys it can only be detected by taking vaginal swabs (Carrol et al., 2007). In the small callitrichid primates, such as marmosets, menstruation is essentially absent, with most of the material reabsorbed rather than lost (Martin, 2003). Menstrual blood loss is observed in our closest relatives, the chimpanzees. Still, it appears that humans are the champions. In addition, human females appear to have a greater number of menstrual periods than other menstruating mammals, including our closest relatives. In young women, a majority of menstrual cycles seem to be anovulatory (Brosens et al., 2009). And, in addition, there is the high rate of implantation failure and early pregnancy loss discussed above, which results in normal-to-heavy menstrual flow, albeit delayed by a few weeks. So not only do women experience greater blood loss through menstruation than our nonhuman relatives, but they experience it more often. The title to this section should perhaps be modified to "Why do human females lose so much blood from menstruation?"

One proposed explanation is that repeat decidualization of the uterine epithelium serves to precondition the uterus for successful implantation (Brosens et al., 2009). As described in chapter 7, successful implantation of a human blastocyst requires extensive remodeling of the uterine epithelium and recruitment of appropriate uNK cells. These preparations are started well before an egg is fertilized and even further ahead of when a blastocyst attempts to implant. Implantation is an inherently inflammatory process, and preconditioning refers to the biological phenomenon that a brief (regulated) exposure to a harmful stimulus protects against tissue damage when a subsequent more acute, unregulated stimulus occurs (Brosens et al., 2009). Thus, several cycles of menstruation may serve to protect the uterus from damage due to the more profound alterations and inflammation of pregnancy.

In the next section, we discuss the emerging perspective of the human decidua as active tissue that engulfs the blastocyst, as opposed to a passive tissue invaded by the blastocyst. Neither conception probably accurately captures the true biological phenomenon. However, it does suggest that human deciduas may have a more extensive and active function in determining embryo quality and in protecting viable embryos from pathogens. The evolved human implantation system, representing the current compromises from the competing maternal-fetal-paternal fitness imperatives, may require more extensive decidual development and a more emphatic reaction to either no pregnancy or a nonviable pregnancy. And, surely, to a certain extent human menstruation simply reflects the more extensive decidual development necessary due to the interstitial and deeply penetrating implantation by the placenta.

Invasion of Uterine Tissue by the Human Placenta

The human blastocyst invades the uterine wall to a greater extent than do the blastocysts of most other anthropoid primates. After implantation, extravillous trophoblast cells invade deeply into the endothelium. This so-called aggressive implantation has been suggested as critical to the development of what is often considered the most essentially human characteristic of our species, our large brain. We discussed the evidence for this theory in chapter 4 and generally concluded that there is little evidence to support the contention that this deep, aggressive implantation has significant benefits in nutrient transport, signaling systems, or even immunoglobulin transport. However, lack of evidence does not doom this theory, though it raises skepticism. In this section, we examine the process of the interstitial implantation of humans (and great apes) in more detail, in order to explore its possible functional and evolutionary significance.

It is interesting that much of the language used in human placental biology is adversarial. Imprinted genes arise from conflict between the genetic actors. The placenta is a battleground for the conflicting fitness imperatives. Extravillous cytotrophoblast erodes/destroys maternal endothelial cells surrounding the spiral arteries. The blastocyst invades the uterine epithelium. In this last example, however, Allen Enders (personal communication) has pointed out that this so-called invasion is not initially particularly destructive. The blastocyst uses adhesion molecules to slide between epithelial cells through existing junctions. Once established under the epithelial layer, the blastocysts do begin to secrete molecules that will destroy maternal epithelial cells, creating a cavity within which the placenta can grow and gain direct access to maternal blood, by eroding all the way into the linings of maternal blood vessels. However, at the site of the initial penetration of uterine epithelium, uterine cells heal over the passageway, creating the decidua capsularis, which consists of a layer of intact maternal epithelial cells between the developing blastocyst and the uterus. One could term the initial process a burrowing under a protective layer of maternal tissue rather than an invasion. And, inarguably, the maternal epithelial layer between the blastocyst and the uterus does provide the blastocyst with protection from pathogens. After development of the syncytiotrophoblast and the amniotic sac, the placenta and developing fetus will have additional membrane layers of protection. But in very early gestation, just after implantation, the blastocyst would be vulnerable. A viable hypothesis is that one adaptive function of the burrowing into the uterine wall by the human blastocyst is to avoid pathogens floating free in the uterus. This, plainly, is beneficial to the fitness of both mother and fetus—an example of cooperation rather than conflict.

Another perspective has been proposed; rather than the blastocyst invading the uterine epithelium, perhaps the more accurate description is that decidual cells engulf and encapsulate the blastocyst (Brosens and Gellersen, 2010). In other words, the decidua is not a passive, receptive tissue awaiting invasion by an active, aggressive blastocyst, but instead the decidua is an active tissue waiting for a blastocyst to attach, at which point it springs into action, engulfing and encapsulating the blastocyst, thereby both sheltering it from potential pathogens and also establishing a controlled environment in which the signaling system can act to determine whether this is a pregnancy to continue or one to be terminated through the natural, cyclical process of menstruation. Interstitial implantation provides the blastocyst with both a safe and well-regulated environment in which to develop. It also, potentially, gives the mother greater ability to signal to and receive signals from the embryo.

The depth of the invasion of maternal epithelium by extravillous trophoblast cells is a regulated phenomenon. There is a complex dance between placental and maternal signals to achieve a successful invasion by extravillous cytotrophoblasts to the tips of the spiral arteries, where they then replace the endothelial cells lining the distal ends of those arteries, transforming them into widely dilated conduits of maternal blood; this process creates a low-resistance, high flow of blood to the maternal-placental interface (Ashton et al., 2005). Insufficient invasion of the decidua by extravillous cytotrophoblasts in humans is associated with intrauterine growth restriction (IUGR) and pre-eclampsia; overly deep invasion causes placenta accreta—a serious condition, because it can lead to severe maternal hemorrhage after birth as the expulsion of the placenta removes more maternal tissue. In the most extreme cases, the trophoblast invasion can extend deeply into the myometrium or even penetrate the uterine wall completely and attach to other organs (termed placenta increta and placenta percreta, respectively). These placental abnormalities are rarer than those caused by too shallow an implantation. This asymmetry makes evolutionary sense; the negative fitness consequences of too deep an invasion by the placenta are more serious than those for too shallow an implantation. The former can cause maternal death due to massive hemorrhaging post partum, whereas the latter only affects the current pregnancy and in some instances may not even be maladaptive (in the reproductive sense) for the fetus. Many of the diseases that small-for-gestational-age babies are at higher risk for are diseases of later adult life and may not directly impact reproduction.

Placental Infections

Infection is a major risk factor for pregnancy complications and poor outcome. It has been associated with spontaneous abortion, preterm labor, and stillbirth. The maternal immune system cannot afford to be truly suppressed, and it is definitely capable of mounting aggressive and potent responses to uterine infections.

Because the fetus is enclosed in the placental membranes, infection of the fetus has to come through the placenta. Pathogens can reach the placenta either by ascending from the lower genital tract through the cervix into the uterus or from maternal blood. A significant proportion of pathogens that infect the placenta have obligate, or at least facultative, intracellular lifecycles. For example, viruses such as cytomegalovirus, protozoa such as *Toxoplasma gondii*, and bacteria such as *Listeria monocytogenes* are all infectious pathogens that invade the placenta by way of maternal blood; all have lifecycles within host cells. At least in the case of *Listeria,* infection of the placenta from maternal blood seems to occur via extravillous cytotrophoblasts located at the villous tips. Notably, the syncytiotrophoblast appears to be resistant to invasion by this pathogen (Robbins et al., 2010); as discussed in chapter 3, it acts as a barrier.

One contribution by Profet from her unlikely hypothesis regarding a pathogen-cleansing function of menstruation was to provide an immunological and adaptive context within which to view menstruation (Howes, 2010). Pathogens in the female reproductive tract, sperm-borne or not, are potentially important selective factors that may have influenced implantation and decidual function. Howes (2010) extended the immunological hypothesis to include a role for menstruation in the resetting of the immunological context of pregnancy. The receptive state of the pregnant uterus has altered both uNK cell populations and signaling that features cooperation with the blastocyst and regulation of maternal tissue development. After menstruation, the uterine immune system returns to a focus on defense against nonself. We still favor considering menstruation (defined as the loss of blood and tissue following a failure to conceive successfully, for whatever reason) a consequence of adaptations (e.g., interstitial implantation) along our lineage and not as a selected-for phenomenon on its own. However, interstitial implantation may have been favored, in part, due to pathogen-defense advantages, and the necessary decidual changes (including immunological factors) to sustain the great ape system of implantation may have resulted in significant menstrual blood loss as a consequence.

Preterm Birth

Preterm labor and birth affects approximately 13% of pregnancies in the United States and is associated with increased neonatal mortality and morbidity, with potential life-long adverse affects. In the United States, African American women have a higher rate of preterm birth (Spong et al., 2011). Although rates of preterm birth for African American mothers have remained stable since 1990 (Muglia and Katz, 2010), they are still more than 50% higher than the rates for non-Hispanic white women (in 2006, 18.4% versus 11.7%). There was a small decline in the preterm birth rate among all U.S. women during 2008, with the rate for non-Hispanic white women decreasing to 11.1% and the rate for African American women dropping to 17.5% (Martin et al., 2009).

Much of the early research concerning preterm birth was focused on improving neonatal outcomes and has been quite successful. However, preterm infants still suffer from higher rates of developmental difficulties. Even worse, the rate of preterm birth continues to increase. There are an estimated 13 million preterm births worldwide every year (Beck et al., 2010), with the highest rates being in Africa, followed by North America.

Preterm birth has long been considered mostly unpredictable. A majority of cases are considered idiopathic, with no obvious causal or risk factors. In large part this lack of understanding of the causes of preterm birth stems from an ignorance of the mechanisms and signals that initiate normal human parturition (Kamel, 2010). It is not understood whether preterm birth results from an acceleration of normal developmental processes, resulting in the uterine environment progressing from the quiescent to the active labor state more rapidly, or represents a pathological condition or a reaction to pathology. The initiation of labor stems from a complex series of biochemical and physiological events that are still only partially defined and understood. Placental signaling certainly has an important place, with increased production of prostaglandins, oxytocin, CRH, cortisol, and inflammatory cytokines all being employed.

Known risk factors exist for spontaneous preterm labor and birth. Infections of the maternal reproductive tract are some examples. Besides ethnic background, maternal factors associated with an increased risk of preterm birth include: smoking, alcohol consumption, substance abuse, low maternal body mass index, maternal age (over 40 or under 18), and a short interpregnancy interval (DeFranco et al., 2007). In one study examining the effects of African American heritage on the risk of preterm birth, more than half of the variability in gestational age was accounted for by environmental factors (York et al., 2010). There is evidence for maternal genome effects but not much evidence for paternal genetic effects (Little, 2009). For example, women whose maternal half-sister had experienced a preterm birth were at greater risk, but this was not true if the sister were a paternal half-sister (Boyd et al., 2009). The risk of a woman experiencing preterm birth increases if she was born preterm, but the father's gestational age at birth does not appear to have an effect on the likelihood of his children being born premature (Wilcox et al., 2008).

Maternal genetic factors accounted for about 14% of the variance in gestational age in both European American and African American women. Interestingly, fetal genetic factors lent a significant contribution to variation in gestational age in European American women (35%) but a negligible one in African American women (York et al., 2010). One study (Palomar et al., 2007) found an association of African American ethnicity with preterm birth for both mother and father, with adjusted odds ratios for extreme preterm birth (<28 weeks gestation) of 1.22–2.14, 1.60–4.20, and 3.33–4.03 for European American mother / African American father, African American mother / European American father, and both parents African American, respectively. Ethnicity is not a helpful marker for genetic variation in general, and the results of York et al. (2010) caution that environmental factors appear to be substantially involved in preterm risk among African Americans. The evidence for paternal genetic factors increasing the risk of preterm birth is weak at best. The evidence that African American heritage is a risk factor is stronger, but the extent to which genetic factors underlie the increased risk is uncertain.

Some authors have suggested adaptive purposes for preterm birth. From the maternal perspective, a compromised fetus may represent a poor investment in time and reproductive resources. Ending the pregnancy would have net positive selective benefits, assuming that future pregnancies might fare better. The earlier the pregnancy could be terminated, the less costly for the mother. However, even losing a fetus in late pregnancy might have been adaptive in the past, in terms of future reproductive success. Women had relatively few reproductive opportunities in the ancient past, as interbirth intervals were likely three or more years. Unquestionably, from the fetal evolutionary perspective, any chance at life is better than none. Late preterm birth, in which there was a chance for survival even in the distant past, might represent a rescue from an inhospitable intrauterine environment, due either to external factors, for example, infection, or to maternal characteristics, such as poor spiral artery remodeling, leading to insufficient blood flow into the intervillous space. Still, the adaptive significance of preterm birth in our past was most likely a complete loss for the fetus and a chance to try again for the mother.

Why are human beings so susceptible to preterm birth? And why is the preterm birth rate so resistant to modern medical and public health interventions? Perhaps preterm birth is one more reflection of the so-called inefficiency of human reproduction. Implantation appears to be inherently inefficient in our species, with significant early fetal loss and poor implantation that leads to preeclampsia and intrauterine growth restriction (see next section). However, the great apes seem to share our implantation characteristics, but they are apparently not susceptible to preterm birth (though they may be to preeclampsia; see below). The interaction of the placental CRH axis with the maternal stress and emotional response systems (hypothalamus-pituitary-adrenal axis, amygdala, and perhaps even CRH receptors in the gut and other peripheral organs) may produce a vulnerability to preterm labor that has arisen due to the human environment changing so dramatically from what it was in our ancestral past. Not that the past was necessarily pleasant, but perhaps the challenges were more predictable and more amenable to physiological and psychosocial coping mechanisms. Modern rates of preterm birth may reflect an interaction of our evolved biology with altered environmental pressures, similar to, though less dramatic than, the obesity epidemic.

Preeclampsia

Preeclampsia is a pregnancy complication that arises in the second half of gestation, after the twentieth week of pregnancy. Hypertensive disorders during pregnancy, including preeclampsia and eclampsia, affect 10% of pregnancies in the United States. The incidence of preeclampsia has been on the rise, but that may relate to improved diagnosis rather than a true increase. Preeclampsia occupies one part of a spectrum of hypertensive disorders in pregnancy. Gestational hypertension, in many senses, is an exaggerated case of the normal changes in metabolism occurring during pregnancy. It is not intrinsically different, just an extreme end of a spectrum of normal physiological changes that occur during gestation. Preeclampsia, however, represents a pathological phenomenon. It is characterized by maternal hypertension and protein in the urine. If the disorder progresses to maternal convulsions, it is termed eclampsia.

Preeclampsia is a major contributor to neonatal and maternal morbidity and mortality among modern humans, and it affects about 4% of pregnancies. Preeclampsia is characterized by a general dysfunction of maternal vascular endothelium (Carter, 2011). Although the etiology of preeclampsia is still uncertain, it is associated with a poorly developed placental-maternal blood supply connection. This has little consequence early in pregnancy, when the resource demand from the fetus is low. In the third trimester, however, the increased resource need of the fetus begins to conflict with the lower available maternal blood flow provided by the compromised implantation. Maternal blood pressure rises in response, attempting to increase blood flow, resulting in gestational hypertension. Other maternal risk factors are important in determining whether this hypertensive disorder progresses into preeclampsia and even more serious conditions such as HELLP syndrome (hemolysis, elevated liver enzymes, low platelet count). Preeclampsia causes changes in the kidney, which result in excess protein being excreted in the urine. In general, these kidney changes will reverse after birth, but recent evidence has suggested that in some cases, the damage may be permanent (Munkhaugen and Vikse, 2009).

Although many authors have asserted that preeclampsia is a uniquely human disease (e.g., Robillard et al., 2003), the evidence suggests that both chimpanzees and gorillas are also susceptible (Carter, 2011; Pijnenborg et al., 2011). Thus, the proposed evolutionary associations of preeclampsia with bipedalism or the large human brain need to be viewed with some skepticism. The interstitial implantation and deep invasion of the uterine lining, with extensive remodeling of the maternal spiral arteries, certainly provides the context in which partial failure of the latter phenomenon leads to the pathology of preeclampsia. But the evolutionary adaptive purpose for this implantation type does not appear to be either bipedalism or a large brain.

Of note, preeclampsia is more common in first-time pregnancies (Roberts and Redman, 1993; Dekker, 2002). It is also more likely to occur when pregnancy occurs after a short period of sexual involvement with the partner (Kho et al., 2009) or if barrier methods of contraception are used (Dekker and Robillard, 2007). A change in partner is also a risk factor, though that situation is usually associated with a long interbirth interval as well, which is also a risk factor. Oral sex with the swallowing of sperm and seminal fluid from the partner is associated with a protective effect (Koelman et al., 2000). These epidemiological findings suggest that preeclampsia has an immune system component and that maternal acquired tolerance to paternal antigens might explain some of the various risk/protective factors that have been identified. This is termed the "immune maladaption hypothesis." As discussed in more detail in chapter 7, implantation is regulated by the maternal immune system, specifically by the action of uterine natural killer (uNK) cells recruited to the decidua and the implantation site. These uNK cells interact with paternal antigens (especially human leukocyte antigen-C) expressed by trophoblasts via killer cell immunoglobulin-like receptors (KIRs). Poor recruitment of uNK cells to the implantation site and/or ineffective signaling via KIRs is hypothesized to compromise the tissue remodeling, leading to a faulty placental connection to the maternal blood supply.

Although the immune maladaption hypothesis has support, both from epidemiological data and through our newer understanding of the role of the maternal uNK cell system in regulating uterine tissue remodeling, preeclampsia appears to be a heterogeneous condition. Multiple paths can lead to preeclampsia, and the condition in young, nulliparous women may differ from that in women with pre-existing cardiovascular disease or other predisposing conditions. There are also several stages of the disease; some authors propose a two-stage model and others a three-stage model. The first stage is a placental stage, involving the invasion of uterine epithelium by trophoblasts and the remodeling of the spiral arteries. With preeclampsia, the invasion is shallower and spiral artery remodeling incomplete. This is the stage of preeclampsia that we focus on here. The other stages are maternal stages and involve maternal physiological reactions to the consequences of stage one. Thus, there are certainly maternal characteristics, including genetic, environmental, and health status, that contribute to the final maternal and infant outcomes deriving from the original abnormality in placentation. Women with pre-existing cardiovascular and metabolic disease are at increased risk of developing preeclampsia. Women who suffer from preeclampsia during a pregnancy are at greater risk for cardiovascular and metabolic diseases later in life. Preeclampsia may be both an indication of underlying vulnerability to disease that is revealed by the added challenges due to pregnancy and also a contributing factor toward a worsening of underlying disease.

Preeclampsia has significant negative effects on maternal and fetal mortality and morbidity and is often associated with intrauterine growth restriction (IUGR). It is difficult to conceive of it arising as an evolutionary adaptation per se. Rather, it appears to be part of the so-called inefficiency of the human reproductive strategy. The deep invasion of trophoblasts into the decidua is a characteristic of the human placenta. Previously, we discussed some of the hypotheses that attempt to explain why this particular maternal-placental connection was favored. Regardless of the selective factors, the evolved process is complicated, with multiple aspects of maternal-placental crosstalk. Preeclampsia may be one part of the spectrum of possibilities, from early pregnancy loss to successful pregnancy, that reflects the complexity and vulnerability of the maternal-placental communication necessary for successful implantation and placental development in our species.

This raises the possibility that the vulnerability to preeclampsia, apparently inherent in the human implantation/placentation process, provided a selective pressure for delaying pregnancy and thus was one of the selective factors underlying the high rate of early pregnancy loss. If the immune maladaption hypothesis is largely correct, or at least is operative in a significant proportion of cases of preeclampsia (and also some cases of IUGR), then repeated exposure to a consistent set of male sperm before pregnancy would be advantageous. It would prime the maternal immune system and lead to better implantation and maternal tissue remodeling, resulting in a better maternal blood supply to the placenta and, on average, better birth outcomes. Being inefficient at getting pregnant is one mechanism to achieve exposure to sperm and seminal fluid for multiple cycles before a successful pregnancy. If menstruation does serve to prime the uterus for more effective decidualization, preconditioning it for implantation, human fertility inefficiency might actually have adaptive value. Perhaps in our distant past, females who experienced many early pregnancy failures actually produced a greater number of successful young over their reproductive lives, generating the offspring who were our ancestors.

Conclusion

Regulation, the ability to control, to greater or lesser extents, the internal environment in the face of external challenges, is a hallmark of life. This facet of life is perhaps most important during early development. Live birth evolved very early in the history of complex, multicellular organisms. We have argued that a primary advantage of live birth is to provide a regulated environment for developing embryos, whether they be seahorse embryos developing in the male's pouch or a human fetus surrounded by an amniotic sac and nourished through an extrafetal organ attached to its mother.

In this book, we have explored the evolution and the evolutionary consequences of the placenta, a fundamental adaptation of the placental mammals. Placental mammals are the descendants of an ancient lineage (the synapsids) that split from the other amniotes more than three hundred million years ago. The only living descendants of this lineage are mammals. The few monotreme mammals still living produce small eggs that are nourished within the reproductive tract (matrotrophic nutrition) for a considerable length of time before they are laid. Marsupials produce even smaller eggs that "hatch" within the reproductive tract and are then nourished for a short time through a choriovitteline placenta (largely histiotrophic nutrition) before they are born in what is almost an embryonic, or at best early fetal, stage. Growth and development are mainly fueled by that other fundamental mammalian adaptation, lactation. The placental mammals have modified the ancient amniotic egg membranes to the greatest extent among mammals, producing a chorio-allantoic placenta that nourishes the embryo and then the fetus, initially via histiotrophic nutrition, but eventually through hemotrophic nutrition, where the gases and nutrients for fetal growth and development are delivered from the maternal bloodstream to the fetus via the placenta. In addition, other molecules with regulatory functions pass to and from the fetus and mother by way of the placenta. The placenta produces, modifies, and regulates many, if not most, of these regulatory signals. The evolution of the placenta greatly extended the ability to regulate the environment in which a mammalian fetus develops.

Although all mammals nourish their offspring after birth with milk, placental mammals have shifted a large proportion of the maternal-to-offspring transfer of resources from lactation to gestation. This change has allowed a greater range of reproductive strategies to evolve, with placental mammal young being born in anything from an extremely altricial state, with eyes and ears closed, as in bears, to offspring that can independently locomote by walking or swimming almost immediately after birth (e.g., giraffes, dolphins). The evolution of a chorio-allantoic placenta was a fundamental successful adaptation at the ancient base of the lineage that eventually produced human beings. The great diversity of placental mammal species is reflected in the diversity of placental structure. Human placental form and function are, truthfully, no more exceptional than that of any of the other placental mammal species. However, it is specific to our lineage, and how and why it has the particular characteristics it does are appropriate scientific questions to explore.

When the placenta is viewed as an organ, the significance of its great diversity of structure becomes even more apparent. Other mammalian organs do not exhibit such morphological variability. Perhaps the brain comes close, but even with all the specializations that have evolved in that organ, the basic mammalian brain is still recognizable across all species. Even the large human brain is similar to other mammalian brains in most aspects. The diversity of placental form implies both great selective pressures acting on placental function but also an inherent flexibility in developmental biology that has allowed placental form to radiate into so many different types. Perhaps most emblematic of this flexibility are the bat species that begin gestation with an endotheliochorial placenta, only to replace it with a newly formed hemochorial placenta at mid-gestation. It would be very exciting if scientists could understand the genetics/epigenetics of that remarkable developmental pattern, where the mammalian placental diversity is on display within an individual animal.

We focused on five themes in this book: the placenta as a regulatory organ; the co-option of the immune system to serve as a regulatory and developmental system; the placenta as a selective force on mechanisms of genetic regulation; the biological anthropology of the placenta; and, in the final chapter, some biomedical aspects of placental biology that relate to the previous four themes. We have taken a regulatory and evolutionary perspective in all we have discussed; regulation is a fundamental aspect of biology, and evolution is a guiding principle for understanding how various aspects of life came about.

The Placenta as a Regulatory/Endocrine Organ

That the placenta functions as an endocrine organ (in addition to having other capacities) and has a potent role in regulating maternal and fetal metabolism is now well accepted. It is part of our modern understanding that all organs have endocrine function and that metabolism and physiology involve to a large extent the coordination of information among organ systems via molecular signals. This in no way downgrades the placenta's role in transporting nutrients and gases from the mother to the fetus and waste products in the reverse direction. It merely recognizes that the placenta is an active participant in this transport system, regulating these processes, and in addition, that growth and especially development rely on a myriad of signals, many of which are not classified as nutrients. Growth factors, cytokines, and other information molecules determine fetal outcome as much as do glucose and oxygen.

The history of knowledge regarding the placenta, as briefly outlined in chapter 1, in many ways simply follows the technological and conceptual advances that have resulted in our modern biological synthesis. Early biological investigations were concerned with anatomy and, as imaging technology improved, progressed to cellular morphology. With the discovery of endocrine secretions, the science of physiology took off, and biology and chemistry became even more tightly interwoven. Since the late twentieth century, our understanding has progressed to where even adipose tissue is considered an endocrine organ. The number and types of molecular signals produced by cells has expanded exponentially, as well. That the placenta has endocrine and regulatory function nowadays appears obvious. The extent to which it actively influences maternal metabolism and fetal growth and development is still being explored; however, the evidence favors a substantial and complex role for placental effects on birth outcome.

The role of the placenta in immunological signaling is perhaps emblematic of our new understanding of the complexity of placental regulatory function. In humans and laboratory rodents, specific types of uterine natural killer cells are recruited to the implantation site through molecular signaling between placental and maternal tissues. These uNK cells are not phagocytotic but rather regulate maternal cellular differentiation, remodeling uterine tissue to be conducive to implantation and placental development. In humans, the uNK cells help to regulate the depth of the extravillous trophoblast invasion of maternal tissue, with insufficient invasion resulting in anything from early pregnancy loss to preeclampsia, with or without intrauterine growth restriction. Too deep an invasion by trophoblasts can result in serious postpartum maternal hemorrhage and possibly maternal death. The placenta, with its signaling, induces an ancient evolved pathogen defense system to act as a collaborator and facilitator to invasion by a foreign body. That foreign body, the blastocyst, is the principal mechanism for reproductive fitness of the mother, so it is unsurprising that placental signaling should result in no attack. That the maternal immune system has evolved to take on regulatory and tissue remodeling functions is fascinating and a particularly important avenue for further scientific exploration.

Genetic Regulation

Epigenetics is arguably the next significant frontier of biological science. The sequencing of the genomes of multiple species, including our own, has shown us both the power of genetics and also our incomplete understanding of how genetic information is regulated. Knowledge of the DNA sequence alone is not as informative as scientists had hoped. As we have consistently argued, understanding regulatory mechanisms is the key to comprehending biological function, and that is true at the genetic level as well. In all species, there are epigenetic mechanisms that act to produce complex organisms with multiple organ systems and cell types. The common DNA sequence of an individual is expressed differently among its cells. Gaining an understanding of how that comes about has long been a prime goal of biologists, ever since the significance of genetic material was grasped. The more we learn, the more we realize how complex and varied the genetic regulatory mechanisms are. Given the extreme diversity of placental form in mammals, the placenta should be fertile ground for exploring how genetic regulatory mechanisms can create different tissues from similar DNA.

The placenta, too, appears to have favored a mechanism of genetic regulation. Gene imprinting, the phenomenon whereby only the maternal or paternal copy of a gene is expressed in the offspring, appears ubiquitous in marsupial and placental mammals but so far is not known to exist in the monotremes, nor in any nonmammalian amniotes. Therefore, gene imprinting would not appear to be an ancient synapsid adaptation (though it is possible it was lost in the monotreme lineage) but instead an adaptation that arose after at least a rudimentary placenta had evolved. Theories proposed by a number of authors, but perhaps most notably by David Haig and his colleagues, explain the evolution of gene imprinting as the result of increased potential for interactive genetic conflict between the fitness goals of the maternal, offspring, and paternal genomes. We suggest that there is also increased potential for cooperative interactions. In either case, the existence of a placenta provided an additional avenue through which selection could act on mechanisms of genetic regulation.

The Human Placenta

Human beings are, not surprisingly, fascinated by our own species' evolution. Understanding our modern biology requires comprehension of the evolutionary pressures and paths that produced us. It is also quite reasonable to hypothesize that changes in one aspect of our biology might be associated with changes in others. Specifically, we examined the hypothesis that the evolution of our large brain required changes in placental biology. This is an extension of the "expensive tissue" hypothesis, which postulates that because brain tissue is metabolically demanding, requiring a higher rate of oxygen and glucose flow compared with other tissue on average, physiological and metabolic adjustments were required in our lineage to compensate.

Many authors have associated the large fetal brain of our offspring with the human pattern of interstitial implantation, including a deeply invasive hemochorial placenta and extensive remodeling of the maternal spiral arteries. Yet, the evidence to date does not support that this form of implantation is required to produce a fetus with a large brain or that these implantation characteristics arose in our lineage alone. To address the first point, many cetacean species (dolphins and whales) produce large-brained calves, despite having an epitheliochorial placenta type. Other species not so well endowed in the brain department nonetheless produce large precocial offspring whose total metabolic requirements at the end of pregnancy are not obviously less than that of a human infant, without having a hemochorial placenta (e.g., horses). Guinea pigs, on the other hand, utilize interstitial implantation and a hemochorial placenta that is at least as deeply invasive as the placenta of human beings. Guinea pigs are wonderful small rodents that produce a small number of precocial babies, quite similar to human reproduction; however, they do not have particularly large brains. So it doesn't appear that a deeply invasive hemochorial placenta is required to produce a large brain or that it is always associated with a large brain. These facts do not negate the hypothesis, as most biological challenges have multiple possible solutions and many biological adaptations "solve" multiple difficulties. Perhaps a careful examination of cetacean placentas will reveal alternative mechanisms to support brain growth, and maybe further examination of guinea pig reproduction will show a different adaptive function for this aspect of its placenta. Although the evidence supports that the human placental bed type is not required to transfer large amounts of maternal resources to the fetus and thus support extensive brain growth, it may still have evolved to perform that function in the human lineage.

However, the evidence from our closest living relatives, the great apes, casts serious doubt on the hypothesis that the human placental type evolved to support the human brain. All great apes employ interstitial implantation (though the lesser apes, the gibbons and siamangs, do not). So with respect to this characteristic, humans do not differ from orangutans, gorillas, or chimpanzees. In addition, both gorillas and chimpanzees possess deeply invading extravillous trophoblast cells, which line the maternal spiral arteries and extensively remodel them to produce a human-like placental bed. The evidence for orangutans is uncertain, but they may resemble human beings in their placental bed morphology as well. Thus, at best this type of placental bed could be said to represent a pre-adaptation that allowed increased brain growth in the human lineage; the evidence suggests that it arose before the divergence of apes and humans and therefore did not evolve because of our increased brain size.

A deeply invasive hemochorial placenta appears to be related to interstitial implantation. For anthropoid primates, that certainly is the case. Most anthropoids, including the smaller ape species, use superficial implantation and a relatively shallow invasion of the uterine epithelium by trophoblast. Only in the great apes and humans do both interstitial implantation and the deep invasion by trophoblast cells occur. To understand the evolution of human placentation, careful consideration of the possible adaptive advantages of interstitial implantation should be undertaken. Perhaps the adaptive significance of human (and great ape) placentation relates more to advantages of interstitial implantation, and the deep invasion of trophoblast is more a consequence of interstitial implantation rather than a trait that was selected for per se.

Does interstitial implantation create a more regulated environment in which an embryo can develop? Interstitial implantation potentially produces an environment that can be better controlled by maternal signaling than can superficial implantation, where the placenta and embryo are exposed to the uterine lumen. With interstitial implantation, the embryo and the placenta are completely surrounded by maternal tissue. Is there evidence that the milieu in which a human or chimpanzee embryo begins its development is any more controlled than the environment of a baboon or macaque embryo?; none of which we are aware. There are differences in placental signaling, however; specifically, the pattern of placental corticotropin-releasing hormone secretion differs between the apes, which utilize interstitial implantation, and the monkeys, which undergo superficial implantation. Is the peak CRH secretion during early-to-mid pregnancy in monkeys related to superficial implantation?—an interesting hypothesis, but one for which there are no published data to evaluate. Data on the pattern of placental CRH secretion in gibbons, apes that employ superficial implantation, would be instructive in judging whether this is a worthwhile hypothesis to pursue.

Interstitial implantation could serve to protect the placenta and embryo against pathogens within the lumen of the uterus. Uterine epithelium provides an additional barrier to pathogens. There certainly is evidence that the human lineage passed through time periods with extensive exposure to pathogens. Human milk and human placental transfer of immunoglobulin G both appear to have changed in response to selection pressures from pathogens attacking fetuses/neonates. However, once again the initial adaptation of interstitial implantation preceded these times in human evolution, making interstitial implantation a pre-adaptation rather than an adaptation. Was there a significant outbreak of sexually transmitted diseases during great ape evolutionary history, prior to the divergence of the orangutan lineage? What alternative advantages of interstitial implantation might exist that were relevant to great ape evolution?

The bottom line regarding the specifics of the human placental bed and the evolution of our large brain is that there is no evidence to support an association. Undoubtedly, our implantation and placentation biology has served our species well; the human placental bed can successfully support the nutrient transfer and endocrine signaling necessary to produce large-brained neonates. But it is not at all clear that the human placental bed differs significantly in these aspects from that of chimpanzees or gorillas. The basic plan predates the human lineage.

Regardless, there are aspects of human reproduction that are unusual, if not peculiar to humans, that derive from placental conditions: high rates of early pregnancy loss, morning sickness, and preeclampsia are some of the examples we explored. We have suggested a possible adaptive link among these conditions, related to the hypothesis that human inefficiency in becoming pregnant is neither pathology nor even an unfortunate by-product of other evolved aspects of our reproductive biology. Instead, we propose that a delay between the commencement of copulations between a female and one (or more) male(s) and pregnancy served an adaptive function for the female.

Human evolutionary history encompasses a long period in which our ancestors most likely lived in what were probably cohesive, small groups. Cooperation was a hallmark of the successful foraging strategy of these ancestors, and cooperation surely extended to many other aspects of life. Females that established consistent, supportive social relationships with other members of their group probably were more reproductively successful, due to the assistance they received in rearing their offspring. These relationships could easily have included a stable sexual relationship with a consistent set of males. A change of males might, on average, represent an unstable and risky situation in which to produce a child. However, if those males were still around six months later, then perhaps stability had been reestablished. Not being in a hurry to get pregnant may have been a highly successful long-term reproductive strategy for our female ancestors.

We propose this hypothesis, admittedly speculative, for a number of reasons. First, we find it interesting, especially in that it links many disparate characteristics of human female reproduction. We hope it stimulates thought and research. Regardless of whether the results of those additional scholarly endeavors advance or discredit the hypothesis, we hope they will advance general knowledge of our evolution and of ourselves as a species. In addition, we offer this hypothesis as a counter to a tendency for evolutionary explanations of human characteristics to focus on efficiency and "progress." Natural selection does not always favor the most efficient traits, nor does it always lead to what humans would consider progress and advancement. Evolution favors those individuals that bequeath more of their DNA to future generations, by whatever means. In a related vein, we submit this hypothesis as a cautionary tale to biomedicine, in which conditions that we as a modern species have deemed undesirable are frequently treated as pathology, without considering that they might represent, or derive from, previously successful adaptations from a time long past that was markedly different from today.

REFERENCES

Adams KM, Yan Z, Stevens AM, Nelson JL. 2007. The changing maternal "self" hypothesis: a mechanism for maternal tolerance of the fetus. Placenta 28: 378–382.

Adkins RM, Nekrutenko A, Li WH. 2001. Bushbaby growth hormone is much more similar to nonprimate growth hormones than to rhesus monkey and human growth hormones. Mol Biol Evol 18: 55–60.

Adler C, Adler EB. 1970. Great men of Padua University, Italy. S Afr Med 44(20): 596–599.

Ahima RS. 2006. Adipose tissue as an endocrine organ. Obesity 14: 242S–249S.

Aiello LC, Wheeler P. 1995. The expensive-tissue hypothesis. Curr Anthropol 36(2): 199–221.

Alwasel SH, Abotalib Z, Aljarallah JS, Osmond C, Alkharaz SM, Alhazza IM, Harrath A, Thornburg K, Barker DJP. 2011. Secular increase in placenta weight in Saudi Arabia. Placenta 32: 391–394.

Ansari TI, Egbor M, Meyers H, Green CJ, Sibbons PD. 2005. Gestational dependent PET placental pathogenesis: defining 2 specific subsets. In Proceedings of a workshop on comparative placentology, ed. PD Sibbons and JF Wade, 33–35. Newmarket, UK: R & W Communications.

Arima T, Hata K, Tanaka S, Kusumi M, Li E, Kato K, Shiota K, Sasaki H, Wake N. 2006. Loss of the maternal imprint in Dnmt3L$^{mat-/-}$ mice leads to a differentiation defect in the extraembryonic tissue. Dev Biol 297(2): 361–373.

Ashton SV, Whitley GSJ, Dash PR, Wareing M, Crocker IP, Baker PN, Cartwright JE. 2005. Uterine spiral artery remodeling involves endothelial apoptosis induced by extravillous trophoblasts through Fas/FasL interactions. Arterioscler Thromb Vasc Biol 25(1): 102–108.

Astwood EB, Greep RO. 1938. A corpus luteum stimulative substance in the rat placenta. Proc Soc Exp Biol Med 38: 713.

Athanassakis I, Farmakiotis V, Aifantis A, Gravanis A, Vassiliadis S. 1999. Expression of corticotrophin-releasing hormone in the mouse uterus: participation in embryo implantation. J Endocrinol 163: 221–227.

Azzarello MY. 1991. Some questions concerning the syngnathidae brood pouch. B Mar Sci 49(3): 741–747.

Bamberger AM, Minas V, Kalantaridou SN, Radde J, Sadeghian H, Löning T, Charalampopoulos I, Brümmer J, Wagener C, Bamberger CM, et al. 2006. Corticotropin-releasing hormone modulates human trophoblast invasion through carcinoembryonic antigen-related cell adhesion molecule-1 regulation. Am J Path 168: 141–150.

Barker DJP. 1991. Fetal and infant origins of adult disease. Br Med J. 301: 1111.

Barr ML, Bertram EG. 1949. A morphological distinction between neurons of the male and female, and the behaviour of the nucleolar satellite during accelerated nucleoprotein synthesis. Nature 163: 676–677.

Bartolomei MS. 2009. Genomic imprinting: employing and avoiding epigenetic processes. Genes Dev 23: 2124–2133.

Basu S, Haghiac M, Surace P, Challier J-C, Guerre-Millo M, Singh K, Waters T, Minium J, Presley L, Catalano PM, et al. 2011. Pregravid obesity associates with increased endotoxemia and metabolic inflammation. Obesity 19: 476–482.

Bauman DE, Curie WB. 1980. Partitioning of nutrients during pregnancy and lactation: a review of mechanisms involving homeostasis and homeorhesis. J Dairy Sci 63: 1514–1529.

Beck S, Wojdyla D, Say L, Betran AP, Merialdi M, Requejo JH, Rubens C, Menon R, Van Look PFA. 2010. The worldwide incidence of preterm birth: a systematic review of maternal mortality and morbidity. Bull World Health Organ 88: 31–38.

Benirschke K, Kaufmann P, Baergen RN. 2006. Pathology of the Human Placenta. New York: Springer.

Bielanska M, Tan SL, Ao A. 2002. High rate of mixoploidy among human blastocysts cultured in vitro. Fertil Steril 78: 1248–1253.

Biensen NJ, Wilson ME, Ford SP. 1998. The impact of either a meishan or Yorkshire uterus on meishan or Yorkshire fetal and placental development to days 70, 90, and 110 of gestation. J Anim Sci 76: 2169–2176.

Blackburn DG. 2006. Squamate reptiles as model organisms for the evolution of viviparity. Herpetol Monogr 20(1): 131–146.

Blackburn P, Wilson G, Moore S. 1977. Ribonuclease inhibitor from human placenta. Purification and properties. J Biol Chem 252: 5904–5910.

Blaise S, de Parseval N, Bénit L, Heidmann T. 2003. Genomewide screening for fusogenic human endogenous retrovirus envelopes identifies syncytin 2, a gene conserved on primate evolution. PNAS, 100: 13013–13018.

Bodnar LM, Siega-Riz AM, Simhan HN, Himes KP, Abrams B. 2010. Severe obesity, gestational weight gain, and adverse birth outcomes. Am J Clin Nutr 91: 1642–1648.

Böhne A, Brunet F, Galiana-Arnoux D, Schultheis C, Volff J-N. 2008. Transposable elements as drivers of genomic and biological diversity in vertebrates. Chromosome Res 16(1): 203–215.

Bonnin A, Goeden N, Chen K, Wilson ML, King J, Shih JC, Blakely RD, Deneris ES, Levitt P. 2011. A transient placental source of serotonin for the fetal forebrain. Nature 472: 347–350.

Bonnin A, Levitt P. 2011. Fetal, maternal, and placental sources of serotonin and new implications for develomental programming of the brain. Neurosci 197: 1–7.

Bowman ME, Lopata A, Jaffe RB, Golos TG, Wickings J, Smith R. 2001. Corticotropin-releasing hormone-binding protein in primates. Am J Primatol 53(3): 123–130.

Boyd HA, Poulsen G, Wohlfahrt J, Murray JC, Feenstra B, Melbye M. 2009. Maternal contributions to preterm delivery. Am J Epidemiol 170: 1358–1364.

Brawand D, Wahli W, Kaessmann H. 2008. Loss of egg yolk genes in mammals and the origin of lactation and placentation. PLoS Biol 6(3): e63.

Breschi MC, Seghieri G, Bartolomei G, Gironi A, Baldi S, Ferrannini E. 1993. Relation of birthweight to maternal glucose and insulin concentrations during normal pregnancy. Diabetologia 36: 1315–1321.

Brosens JJ, Gellersen B. 2010. Something new about early pregnancy: deciding biosensoring and natural embryo selection. Ultrasound Obstet Gynecol 36: 1–5.

Brosens JJ, Parker MG, McIndoe A, Pijnenborg R, Brosens IA. 2009. A role for menstruation in preconditioning the uterus for successful pregnancy. Am J Obstet Gynecol 200: 615.e1–615e.6.

Broussard CN, Richter JE. 1998. Nausea and vomiting of pregnancy. Gastroenterol Clin North Am 27: 123–151.

Brown CJ, Greally JM. 2003. A stain upon the silence: genes escaping X inactivation. Trends Genet 19: 432–438.

Brown CJ, Lafreniere RG, Powers VE, Sebastio G, Ballabio A, Pettigrew AL, Ledbetter DH, Levy E, Craig IW, Willard HF. 1991. Localization of the X inactivation centre on the human X chromosome in Xq13. Nature 349: 82–84.

Burkart JM, Hrdy SB, Van Sckaik CP. 2009. Cooperative breeding and human cognitive evolution. Evol Anthropol 18: 175–186.

Burton GJ, Jauniaux E, Charnock-Jones DS. 2010. The influence of the intrauterine environment on human placental development. Int J Dev Biol 54: 303–311.

Burton GJ, Watson AL, Hempstock J, Skepper JN, Jauniaux E. 2002. Uterine glands provide histiotrophic nutrition for the human fetus during the first trimester of pregnancy. J Clin Endocrinol Metab 87(6): 2954–2959.

Butts CL, Candando KM, Warfel J, Belyavskaya E, D'Agnillo F, Sternberg EM. 2010. Progesterone regulation of uterine dendritic cell function in rodents is dependent on the stage of the estrous cycle. Mucosal Immunology 3: 496–505.

Campbell FM, Gordon MJ, Dutta-Roy AK. 1998. Placental membrane fatty acid–binding protein preferentially binds arachidonic and docosahexaenoic acids. Life Sci 63: 235–240.

Cannon WB. 1935. The stresses and strains of homeostasis. Am J Med Soc 189: 1–14.

Carrel L, Park C, Tyekucheva S, Dunn J, Chiaromonte F, Makova KD. 2006. Genomic environment predicts expression patterns on the human inactive X chromosome. PLoS Genet 2(9): e151.

Carrol RL, Mah K, Fanton JW, Maginnis GN, Brenner RM, Slayden OD. 2007. Assessment of menstruation in the vervet (*Cercopithecus aethiops*). Am J Primatol 69: 901–916.

Carter AM. 1999. J. P. Hill on placentation in primates. Placenta 20: 513–517.

Carter AM. 2011. Comparative studies of placentation and immunology in non-human primates suggest a scenario for the evolution of deep trophoblast invasion and an explanation for human pregnancy disorders. Reprod 141: 391–396.

Carter AM, Enders AC. 2004. Comparative aspects of trophoblast development and placentation. Rep Biol Endocrinol 2: 46–60.

Carter AM, Mess A. 2010. Hans Strahl's pioneering studies in comparative placentation. Placenta 31: 848–852.

Carter AM, Pijnenborg R. 2010. Evolution of invasive placentation with special reference to non-human primates. Best Pract Res Cl Ob 25: 249–257.

Castracane VD, Henson MC. 2002. When did leptin become a reproductive hormone? Semin Reprod Med 20: 89–92.

Cerruti RA, Lyons WR. 1960. Mammogenic activities of the midgestational mouse placenta. Endocrinology 67: 884–887.

Challier JC, Basu S, Bintein T, Minium J, Hotmire K, Catalano PM, Haugel-de Mouzon S. 2008. Obesity in pregnancy stimulates macrophage accumulation and inflammation in the placenta. Placenta 29: 274–281.

Challis JR, Lockwood CJ, Myatt L, Norman JE, Strauss JF, Petraglia F. 2009. Inflammation and pregnancy. Reprod Sci 16(2): 206–215.

Challis J, Sloboda D, Matthews S, Holloway A, Alfaidy N, Howe D, Fraser M, Newnham J. 2000. Fetal hypothalamic-pituitary adrenal (HPA) development and activation as a determinant of the timing of birth, and of postnatal disease. Endocr Res 26(4): 489–504.

Chan T-F, Su J-H, Chung Y-F, Hsu Y-H, Yeh Y-T, Yuan S-SF. 2006. Elevated amniotic fluid leptin levels in pregnant women who are destined to develop preeclampsia. Acta Obstet Gynecol 85: 171–174.

Chang CL, Hsu SY. 2004. Ancient evolution of stress-regulating peptides in vertebrates. Peptides 25: 1681–1688.

Chaouat G. 1999. Regulation of T-cell activities at the feto-placental interface—by placenta? Am J Reprod Immunol 42: 199–204.

Chaumeil J, Le Baccon P, Wutz A, Heard E. 2006. A novel role for Xist RNA in the formation of a repressive nuclear compartment into which genes are recruited when silenced. Genes Dev 20: 2223–2237.

Chu SY, Kim SY, Lau J, Schmid CH, Dietz PM, Callaghan WM, Curtis KM. 2007. Maternal obesity and risk of stillbirth: a metaanalysis. Am J Obstet Gynecol 197: 223–228.

Clifton VL, Telfer JF, Thompson AJ, Cameron IT, Teoh TG, Lye SJ, Challis JRG. 1998. Corticotropin-releasing hormone and proopiomelanocortin-derived peptides are present in human myometrium. J Clin Endocrinol Metab 83(10): 3716–3721.

Clinton, H. 1996. It Takes A Village: And Other Lessons Children Teach Us. New York: Simon & Schuster.

Coe CL, Lubach GR, Izard KM. 1994. Progressive improvement in the transfer of maternal antibody across the order Primates. Am J Primatol 32: 51–55.

Cole LA. 2007. Hyperglycosylated hCG. Placenta 28: 977–986.

Cole LA. 2009. hCG and hyperglycosylated hCG in the establishment and evolution of hemochorial placentation. J Reprod Immunol 82: 112–118.

Comfort N. 2001. The Tangled Field. Harvard University Press.

Conley AJ, Pattison JC, Bird IM. 2004. Variations in adrenal androgen production among (nonhuman) primates. Semin Reprod Med 22: 311–326.

Creighton HB, McClintock B. 1931. A correlation of cytological and genetical crossing-over in *Zea mays*. PNAS 17: 492–497.

Crews D. 2003. Sex determination: where environment and genetics meet. Evol Dev 5: 50–55.

Crews D, McLachlan JA. 2006. Epigenetics, evolution, endocrine disruption, health and disease. Endocrinology 147(6): s4–s10.

Crofton Long E. 1963. The placenta in lore and legend. Bull Med Libr Assoc 51(2): 233–241.

Cubas P, Vincent C, Coen E. 1999. An epigenetic mutation responsible for natural variation in floral symmetry. Nature 401: 157–161.

Dallman MF, Akana SF, Strack AM, Hanson ES, Sebastian RJ. 1995. The neural network that regulates energy balance is responsive to glucocorticoids and insulin and also regulates HPA axis responsivity at a site proximal to CRF neurons. Ann NY Acad Sci 771: 730–742.

Damjanovic SS, Stojic RV, Lalic NM, Jotic AZ, Macut DP, Ognjanovic SI, Petakov MS, Popovic BM. 2009. Relationship between basal metabolic rate and cortisol secretion throughout pregnancy. Endocrine 35: 262–268.

Davies W, Isles AR, Burgoyne PS, Wilkinson LS. 2006. X-linked imprinting: effects on brain and behavior. BioEssays 28: 35–44.

Davies W, Isles AR, Wilkinson LS. 2005. Imprinted gene expression in the brain. Neurosci Biobehav Rev 29: 421–430.

Davis M. 2004. Nausea and vomiting of pregnancy: an evidence-based review. J Perinat Neonatal Nurs 18: 312–328.

Deakin JE, Chaumeil J, Hore TA, Graves JAM. 2009. Unravelling the evolutionary origins of X chromosome inactivation in mammals: insights from marsupials and monotremes. Chromosome Res 17(5): 671–685.

Deanesly R, Newton WH. 1940. The influence of the placenta on the corpus luteum of pregnancy in the mouse. J Endocrinol 2: 317–321.

De Boo HA, Harding JE. 2006. The developmental origins of adult disease (Barker) hypothesis. Austral New Zeal J Obstet Gynaecol 46: 4–14.

DeFranco EA, Stamilio DM, Boslaugh SE, Gross GA, Muglia L. 2007. A short interpregnancy interval is a risk factor for preterm birth and its recurrence. Am J Obstet Gynecol 197: 264.e1–264.e6.

Dekker GA. 2002. The partner's role in the etiology of preeclampsia. J Reprod Immunol 57, 203–215.

Dekker G, Robillard P-Y. 2007. Pre-eclampsia: is the immune maladaptation hypothesis still standing? An epidemiological update. J Reprod Immunol 76: 8–16.

De Mello JCM, de Araújo ESS, Stabellini R, Fraga AM, de Souza JES, Sumita DR, Camargo AA, Pereira LV. 2010. Random X inactivation and extensive mosaicism in human placenta revealed by analysis of allele-specific gene expression along the X chromosome. PLoS ONE 5(6): e10947.

Denison FC, Roberts KA, Barr SM, Norman JE. 2010. Obesity, pregnancy, inflammation, and vascular function. Reprod 140: 373–385.

Denney JM, Nelson EL, Wadhwa PD, Waters TP, Mathew L, Chung EK, Goldenberg RL, Culhane JF. 2011. Longitudinal modulation of immune system cytokine profile during pregnancy. Cytokine 53(2): 170–177.

De Witt F. 1959. An historical study on theories of the placenta to 1900. J Hist Med Allied Sci 14(7): 360–374.

Dindot SV, Kent KC, Evers B, Loskutoff N, Womack J, Piedrahita JA. 2004. Conservation of genomic imprinting at the XIST, IGF2, and GTL2 loci in the bovine. Mamm Genome 15: 966–974.

Donaldson WL, Oriol JG, Plavin A, Antczak DF. 1992. Developmental regulation of class I major histocompatability complex antigen expression by equine trophoblastic cells. Differentiation 52: 69–78.

Dotzler SA, Digeronimo RJ, Roder BA, Silger-Khodr TM. 2004. Distribution of corticotropin releasing hormone in the fetus, newborn, juvenile, and adult baboon. Pediatr Res 55: 120–125.

Dulvy NK, Reynolds JD. 1997. Evolutionary transitions among egg-laying, live-bearing and maternal inputs in sharks and rays. Proc R Soc B 264(1386): 1309–1315.

Dunn PM. 1990. Dr. William Harvey (1578–1657): physician, obstetrician, and fetal physiologist. Arch Dis Child 65: 1098–1100.

Dunn PM. 2003. Dr. Erasmus Darwin (1731–1802) of Lichfield and placental respiration. Arch Dis Child Fetal Neonatal Ed 88: F346–F348.

Dupressoir A, Marceau G, Vernochet C, Bénit L, Kanellopoulos C, Sapin V, Heidmann T. 2005. Syncytin-A and syncytin-B, two fusogenic placenta-specific murine envelope genes of retroviral origin conserved in Muridae. PNAS 102(3): 725–730.

Duttaroy AK. 2009. Transport of fatty acids across the human placenta: a review. Prog Lipid Res 48: 52–61.

Elliot MG, Crespi BJ. 2009. Phylogenetic evidence for early hemochorial placentation in Eutheria. Placenta 30: 949–967.

Ellegren H. 2002. Dosage compensation: do birds do it as well? Trends Genet 18: 25–28.

Emanuel RL, Torday JS, Asokananthan N, Sunday ME. 2000. Direct effects of corticotropin-releasing hormone and thyrotropin-releasing hormone on fetal lung explants. Peptides 21(12): 1819–1829.

Enders AC. 1960. Electron microscopic observations on the villous haemochorial placenta of the nine-banded armadillo. J Anat 94: 205–215.

Enders AC. 1965. A comparative study of the fine structure of the trophoblast in several hemochorial placentas. Am J Anat 116: 29–67.

Enders AC. 1989. Trophoblast differentiation during the transition from the trophoblastic plate to lacunar stage of implantation in the rhesus monkey and human. Am J Anat 186(1): 85–98.

Enders AC. 2002. Implantation in the nine-banded armadillo: how does a single blastocyst form four embryos? Placenta 23(1): 71–85.

Enders AC, Blankenship TN, Conley AJ, Jones CJP. 2006. Structure of the midterm placenta of the spotted hyena, *Crocuta crocuta*, with emphasis on the diverse hemophagous regions. Cells Tissues Organs 183: 141–155.

Enders AC, Carter AM. 2004. What can comparative studies of placental structure tell us?—a review. Placenta 25: S3–S9.

Enders AC, King BF. 1991. Early stages of trophoblastic invasion of the maternal vascular system during implantation in the macaque and baboon. Am J Anat 192(4): 329–346.

Enders AC, Lopata A. 1999. Implantation in the marmoset monkey: expansion of the early implantation site. Anat Rec 256(3): 279–299.

Englund JA. 2007. The influence of maternal immunization on infant immune responses. J Comp Pathol 137(1): S16–S19.

Erickson K, Throsen P, Chrousos G, Grigoriadis DE, Khongsaly O, McGregor J, Schulkin J. 2001. Preterm birth: associated neuroendocrine, medical, and behavioral risk factors. J Clin Endocrinol Metab 86(6): 2544–2552.

Ezaz T, Stiglec R, Veyrunes F, Graves JAM. 2006. Relationships between vertebrate ZW and XY sec chromosome systems. Curr Biol 16(17): R736–R743.

Fain JN. 2006. Release of interleukins and other inflammatory cytokines by human adipose tissue is enhanced by obesity and primarily due to the nonfat cells. Vit Hormones 74: 443–477.

Farley DM, Dudley CDJ, Li C, Jenkins SL, Myatt L, Nathanielsz PM. 2010. Placental amino acid transport and placental leptin resistance in pregnancies complicated by maternal obesity. Placenta 31: 718–724.

Farley D, Tejero ME, Comuzzie AG, Higgins PB, Cox L, Werner SL, Jenkins SL, Li C, Choi J, Dick EJ Jr, et al. 2009. Feto-placental adaptations to maternal obesity in the baboon. Placenta 30: 752–60.

Fasting MH, Oken E, Mantzoros CS, Rich-Edwards JW, Majzoub JA, Kleinman K, Rifas-Shiman SL, Vik T, Gillman MW. 2009. Maternal levels of corticotropin-releasing hormone during pregnancy in relation to adiponectin and leptin in early childhood. J Clin Endocrinol Metab 94: 1409–1415.

Finn CA. 1998. Menstruation: a nonadaptive consequence of uterine evolution. Q Rev Biol 73: 163–173.

Flaxman SM, Sherman PW. 2008. Morning sickness: adaptive cause or nonadaptive consequence of embryo vitality? Am Nat 172: 54–62.

Flexner LB, Roberts RB. 1939. The measurement of placental permeability with radioactive sodium: the relation of placental permeability to fetal size in the rat. Am J Physiol 128: 154–158.

Florio P, Severi FM, Ciarmela P, Fiore G, Calonaci G, Merola A, De Felice C, Palumbo M, Petraglia F. 2002. Placental stress factors and maternal-fetal adaptive response: the corticotrophin-releasing factor family. Endocrine 19(1): 91–102.

Forbes TR. 1953. The social history of the caul. In The Manner Born: Birth Rites in Cross-Cultural Perspective, ed. L Dundes, 119–132.

Forsdahl A. 1977. Are poor living conditions in childhood and adolescence important risk factors for arteiosclerotic heart disease? Br J Prev Soc Med 31: 91–95.

Forssmann WG, Hock D, Lottspeich F, Henschen A, Kreye V, Christmann M, Reinecke M, Metz J, Carlquist M, Mutt V. 1983. The right auricle of the heart is an endocrine organ. Cardiodilatin as a peptide hormone candidate. Anat Embryol (Berl) 168(3): 307–313.

Fox E, Ridgewell A, Ashwin C. 2009. Looking on the bright side: biased attention and the human serotonin transporter gene. Proc R Soc B 276: 1747–1751.

Frazer JG. 1890. The Golden Bough: A Study in Comparative Religion. London and New York: Macmillan.

Frim DM, Emanuel RL, Robinson BG, Smas CM, Adler GK, Majzoub JA. 1988. Characterization and gestational regulation of corticotrophin-releasing hormone messenger RNA in human placenta. J Clin Invest 82(1): 287–292.

Fuke C, Shimabukuro M, Petronis A, Sugimoto J, Oda T, Miura K, Miyazaki T, Ogura C, Okazaki Y, Jinno Y. 2004. Age related changes in 5-methylcytosine content in human peripheral leukocytes and placentas: an HPLC-based study. Ann Hum Genet 68(3): 196–204.

Gazmararian JA, Peterson R, Jamieson DJ, Schild L, Adams MM, Deshpande AD, Franks AL. 2002. Hospitalizations during pregnancy among managed care enrollees. Obstet Gynecol 100(1): 94–100.

Ghirardini G. 1982. Embryology and obstetrics in the work of Columbus Realdus. J Obstet Gynaecol 3(1): 41–42.

Gillman MW, Rich-Edwards JW, Huh S, Majzoub JA, Oken E, Taveras EM, Rifas-Shiman SL. 2006. Maternal corticotropin-releasing hormone levels during pregnancy and offspring adiposity. Obesity 14: 1647–1653.

Gingerich PD. 2005. Cetacea. In Placental Mammals: Origin, Timing, and Relationships of the Major Extant Clades, ed. KD Rose and JD Archibald, 234–252.

Gjerdingen DK, Yawn BP. 2007. Postpartum depression screening: importance, methods, barriers, and recommendations for practice. J Am Board Fam Med 20(3): 280–288.

Glinoer D, De Nayer P, Bourdoux P, Lemone M, Robyn C, Van Steirteghem A, Kinthaert J, Lejeune B. 1990. Regulation of maternal thyroid during pregnancy. J Clin Endocrinol Metab 71(2): 276–287.

Goland RS, Conwell IM, Jozak S. 1995. The effect of pre-eclampsia on human placental corticotrophin-releasing hormone content and processing. Placenta 16(4): 375–382.

Goland RS, Jozak S, Warren WB, Conwell IM, Start RI, Tropper PJ. 1993. Elevated levels of umbilical cord plasma corticotrophin-releasing hormone in growth-retarded fetuses. J Clin Endocrinol Metab 77: 1174–1179.

Goland RS, Wardlaw SL, Fortman JD. 1992. Plasma corticotropin-releasing factor concentrations in the baboon during pregnancy. Endocrinology 131: 1782–1786.

Goland RS, Wardlaw SL, Stark RI, Brown LSJ, Frantz AG. 1986. High levels of corticotropin-releasing hormone immunoreactivity in maternal and fetal plasma during pregnancy. J Clin Endocrinol Metab 63: 1199–1203.

Goldman-Wohl D, Yagel S. 2009. Preeclampsia—a placenta developmental biology perspective. J Reprod Immunol 82: 96–99.

Golos TG, Durning M, Fisher JM, Fowler PD. 1993. Cloning of four growth hormone / chorionic somatomammotropin-related complementary deoxyribonucleic acids differentially expressed during pregnancy in the rhesus monkey placenta. Endocrinology 133: 1744–1752.

Göth A, Booth DT. 2004. Temperature-dependent sex ratio in a bird. Biol Lett 1(1): 31–33.

Grammatopoulos D, Dai Y, Chen J, Karteris E, Papadopoulou N, Easton AJ, Hillhouse EW. 1998. Human corticotrophin-releasing hormone receptor: differences in subtype expression between pregnant and nonpregnant myometria. J Clin Endocrinol Metab 83(7): 2539–2544.

Graves JAM. 2006. Sex chromosome specialization and degeneration in mammals. Cell 124: 901–914.

Graves JAM. 2008. Weird animal genomes and the evolution of vertebrate sex and sex chromosomes. Annu Rev Genet 42: 565–586.

Graves JAM. 2010. Review: sex chromosome evolution and the expression of sex-specific genes in the placenta. Placenta 24: S27–S32.

Graves JAM, Gecz J, Hameister H. 2002. Evolution of the human X—a smart and sexy chromosome that controls speciation and gene. Nature 326: 501–505.

Gregory TR. 2002. A bird's-eye view of the C-value enigma: genome size, cell size, and metabolic rate in the class Aves. Evolution 56, 121–130.

Grino M, Chrousos GP, Margioris AN. 1987. The corticotrophin releasing hormone gene is expressed in human placenta. Biochem Biophys Res Commun 148(3): 1208–1214.

Grosser O. 1909. Vergleichende anatomie und entwicklungsgeschichte der eihaute und der placenta. Wien and Leipzig: W. Braumüller.

Grosser O. 1927. Frühentwicklung, eihautbildung und placentation des menschen und der säugetiere. München: Bergmann.

Haddow JE, McClain MR, Lambert-Messerlian G, Palomaki GE, Canick JA, Cleary-Goldman J, Malone FD, Porter TF, Nyberg DA, Bernstein P, et al. 2008. Variability in thyroid-stimulating hormone suppression by human chorionic gonadotropin during early pregnancy. J Clin Endocrinol Metab 93(9): 3341–3347.

Haggarty P, Ashton J, Joynson M, Abramovich DR, Page K. 1999. Effect of maternal polyunsaturated fatty acid concentration on transport by the human placenta. Biol Neonate 75: 350–359.

Haig D. 1993. Genetic conflicts in human pregnancy. Q Rev Biol 68(4): 495–532.

Haig D. 2006. Self-imposed silence: parental antagonism and the evolution of X-chromosome inactivation. Evolution 60: 440–447.

Haig D. 2008. Placental growth hormone–related proteins and prolactin-related proteins. Placenta 22 Suppl A: S36–S41.

Hallast et al. 2008. High divergence in primate-specific duplicated regions: human and chimpanzee Chorionic Gonadotropin Beta genes. BMC Evol Biol 8: 195.

Hammett FS. 1918. The effect of the maternal ingestion of desiccated placenta upon the rate of growth of breast-fed infants. J Biol Chem 36: 569–573.

Hammett FS, McNeile LG. 1917. Concerning the effect of ingested placenta on the growth-promoting properties of human milk. Science 46: 345–346.

Hannan NJ, Salamonsen LA. 2007. Role of chemokines in the endometrium and in embryo implantation. Curr Opin Obstet Gynecol 19: 266–272.

Harvey W. 1847. The Works of William Harvey. Trans. R Willis. London: Sydenham Society, 1–624.

Hauguel–de Mouzon S, Guerre-Millo M. 2006. The placenta cytokine network and inflammatory signals. Placenta 27: 794–798.

Heard E, Carrel L. 2009. Foreward: coping with sex chromosome imbalance. Chromosome Res 17: 579–583.

Heerwagen MJ, Miller MR, Barbour LA, Friedman JE. 2010. Maternal obesity and fetal metabolic programming: a fertile epigenetic soil. Am J Physiol 299: R711–R722.

Heidmann O, Vernochet C, Dupressoir, Heidmann T. 2009. Identification of an endogenous retroviral envelope gene with fusogenic activity and placenta-specific expression in the rabbit: a new "syncytin" in a third order of mammals. Retrovirology 6: 107–117.

Hendrie TA, Peterson PE, Short JJ, Tarantal AF, Rothgarn E, Hendrie MI, Hendrickx AG. 1996. Frequency of prenatal loss in a macaque breeding colony. Am J Primatol 40: 41–53.

Henke A, Luetjens CM, Simoni M, Grommoll J. 2007. Chorionic gonadotropin beta subunit gene expression in the marmoset pituitary is controlled by steroidogenic factor 1 (SF1), early growth response protein 1 (Egr1) and pituitary homeobox factor 1 (Pitx1). Endocrinology 148: 6062–6072.

Henson MC, Castracane VD. 2002. Leptin: roles and regulation in primate pregnancy. Semin Reprod Med 20: 113–122.

Henson MC, Swan KF, Edwards DE, Hoyle GW, Purcell J, Castracane VD. 2004. Leptin receptor expression in fetal lung increases in late gestation in the baboon: a model for human pregnancy. Reprod 127: 87–94.

Herrmann TS, Siega-Riz AM, Hobel CJ, Aurora C, Dunkel-Schetter C. 2001. Prolonged periods without food intake during pregnancy increase risk for elevated maternal corticotrophin-releasing hormone concentrations. Am J Obstet Gynecol 185(2): 403–412.

Hillier LW, Miller W, Birney E, Warren W, Hardison RC, Ponting CP, Bork P, Burt DW, Groenen MAM, Delaney ME, et al. 2004. Sequence and comparative analysis of the chicken genome provide unique perspectives on vertebrate evolution. Nature 432: 695–716.

Hobel CJ, Dunkel-Schetter C, Roesch SC, Castro LC, Arora CP. 1999. Maternal plasma corticotropin-releasing hormone associated with stress at 20 weeks' gestation in pregnancies ending in preterm delivery. Am J Obstet Gynecol 180: 257–263.

Hobfoll SE, Ritter C, Lavin J, Hulsizer MR, Cameron RP. 1995. Depression prevalence and incidence among inner-city pregnant and postpartum women. J Consult Clin Psych 63(3): 445–453.

Hobson WC, Graham CE, Rowell TJ. 1991. National chimpanzee breeding program: primate research institute. Am J Primatol 24: 257–263.

Hogan MC, Foreman KJ, Najhavi M, Ahn SY, Wang M, Makela SM, Lopez AD, Lozano R, Murray CJL. 2010. Maternal mortality for 181 countries, 1980–2008: a systematic analysis of progress towards Millenium Development Goal 5. Lancet 375: 1609–1623.

Holliday R. 1990. Mechanisms for the control of gene activity during development. Biol Rev 65(4): 431–471.

Holliday R. 2006. Epigenetics: a historical overview. Epigenetics 1(2): 76–80.

Hook EB. 1976. Changes in tobacco smoking and ingestion of alcohol and caffeinated beverages during early pregnancy: are these consequences, in part, of feto-protective mechanisms diminishing maternal exposure to embryotoxins? In Kelly S, Hook EB, Janrich DT, Porter IH, editors. Birth defects: risks and consequences. New York: Academic Press, 173–183.

Hore TA, Koina E, Wakefield MJ, Graves JAM. 2007. The region homologous to the X-chromosome inactivation centre has been disrupted in marsupial and monotreme mammals. Chromosome Res 15: 147–161.

Howe DC, Gertler A, Challis JR. 2002. The late gestation increase in circulating ACTH and cortisol in the fetal sheep is suppressed by intracerebroventricular infusion of recombinant ovine leptin. J Endocrinol 174: 259–266.

Howes M. 2010. Menstrual function, menstrual suppression, and the immunology of the human female reproductive tract. Perspect Biol Med 53: 16–30.

Hrdy SB. 2009. Mothers and Others: The Evolutionary Origins of Mutual Understanding. Cambridge, MA: Belknap Press / Harvard University Press.

Hu D, Cross JC. 2010. Development and function of trophoblast giant cells in the rodent placenta. Int J Dev Biol 54(2–3): 341–254.

Hughes AL, Friedman R. 2008. Genome size reduction in the chicken has involved massive loss of ancestral protein-coding genes. Mol Biol Evol 25: 2681–2688.

Hughes AL, Piontkivska H. 2005. DNA repeat arrays in chicken and human genomes and the adaptive evolution of avian genome size. BMC Evol Biol 5: 12.

Huising MO, Flik G. 2005. The remarkable conservation of corticotropin-releasing hormone (CRH)-binding protein in the honeybee (*Apis mellifera*) dates the CRH system to a common ancestor of insects and vertebrates. Endocrinol 146: 2165–2170.

Huppertz B. 2010. Biology of the placental syncytiotrophoblast—myths and facts. Placenta 31: S75–S81.

Huxley RR. 2000. Nausea and vomiting in early pregnancy: its role in placental development. Obstet Gynecol 95(5): 779–782.

Hyunh KD, Lee JT. 2005. X-chromosome inactivation: a hypothesis linking ontogeny and phylogeny. Nat Rev Genet 6: 410–418.

Ilias I, Mastorakos G. 2003. The emerging role of peripheral corticotropin-releasing hormone (CRH). J Endocrinol Invest 26(4): 364–371.

Itoh Y, Melamed E, Yang X, Kampf K, Wang S, Yehya N, Van Nas A, Replogle K, Band MR, Clayton DF, et al. 2007. Dosage compensation is less effective in birds than in mammals. J Biol 6: 2.

Jaffe RB, Lee PA, Midgley AR Jr. 1969. Serum gonadotropins before, at the inception of, and following human pregnancy. J Clin Endocrinol Metab 29: 1281–1283.

Jakimiuk AJ, Skalba P, Huterski R, Haczynski J, Magoffin DA. 2003. Leptin messenger ribonucleic acid (mRNA) content in the human placenta at term: relationship to levels of leptin in cord blood and placental weight. Gynecol Endocrinol 17: 311–316.

Jauniaux E, Gulbis B, Burton GJ. 2003. The human first trimester gestational sac limits rather than facilitates oxygen transfer to the foetus—a review. Placenta 24: S86–S93.

Jauniaux E, Poston L, Burton GJ. 2006. Placental-related diseases of pregnancy: involvement of oxidative stress and implications in human evolution. Hum Reprod Update 12: 747–755.

Ji Q, Luo Z-X, Yuan C-X, Tabrum AR. 2006. A swimming mammaliaform from the Middle Jurassic and ecomorphological diversification of early mammals. Science 311(5764): 1123–1127.

Ji Q, Luo Z-X, Yuan C-X, Wible JR, Zhang J-P, Georgi JA. 2002. The earliest known eutherian mammal. Nature 416: 816–822.

Jones E, Kay MA. 2003. The Cultural Anthropology of the Placenta. In The Manner Born, ed. L Dundes, 101–116.

Jones PA, Takai D. 2001. The role of DNA methylation in mammalian epigenetics. Science 293(5532): 1068–1070.

Jones SA, Brooks AN, Challis JRG. 1989. Steroids modulate corticotrophin-releasing hormone production in human fetal membranes and placenta. J Clin Endocrinol Metab 68(4): 825–830.

Kalantaridou SN, Makrigiannakis A, Zoumakis E, Chrousos GP. 2004a. Reproductive functions of corticotropin-releasing hormone. Research and potential clinical utility of antalarmins (CRH receptor type 1 antagonists). Am J Reprod Immunol 51: 269–274.

Kalantaridou SN, Makrigiannakis A, Zoumakis E, Chrousos GP. 2004b. Stress and the female reproductive system. J Reprod Immunol 62: 61–68.

Kalantaridou SN, Zoumakis E, Makrigiannakis A, Lavasidis LG, Vrekoussis T, Chrousos GP. 2010. Corticotropin-releasing hormone, stress and human reproduction: an update. J Reprod Immunol 85: 33–39.

Kallen CB. 2004. Steroid hormone synthesis in pregnancy. Obstet Gynecol Clin 31(4): 795–816.

Kamel RM. 2010. The onset of human parturition. Arch Gynecol Obstet 281: 975–982.

Kamo SL, Czamanske GK, Amelin Y, Fedorenko VA, Davis DW, Trofimov VR. 2003. Rapid eruption of Siberian flood-volcanic rocks and evidence for coincidence with the Permian-Triassic boundary and mass extinction at 251 Ma. Earth Planet Sc Lett 214(1–2): 75–91.

Karalis K, Goodwin G, Majzoub JA. 1996. Cortisol blockade of progesterone: a possible molecular mechanism involved in the initiation of human labor. Nat Med 2: 556–560.

Karanth KP, Delefosse T, Rakotosamimanana B, Parsons TJ, Yoder AD. 2005. Ancient DNA from giant extinct lemurs confirms single origin of Malagasy primates. PNAS 102: 5090–5095.

Karteris E, Papadopoulou N, Grammatopoulos DK, Hillhouse EW. 2004. Expression and signaling characteristics of the corticotrophin-releasing hormone receptors during the implantation phase in the human endometrium. J Mol Endocrinol 32(1): 21–32.

Kaufmann P. 1983. Vergleichend-anatomische und funktionelle Aspekte des Placenta-Baues. Funkt Biol Med 2: 71–79.

Kaufmann P. 1992. Classics revisited: Otto Grosser's monographs. Placenta 13: 191–193.

Kawai A, Nishida-Umehara C, Ishikima J, Tsuda Y, Ota H, Matsuda Y. 2007. Different origins of bird and reptile sex chromosomes inferred from comparative mapping of chicken Z-linked genes. Cytogenet Genome Res 117: 92–102.

Keller EF. 1983. A feeling for the organism: the life and times of Barbara McClintock. San Francisco: WH Freeman.

Kershaw EE, Flier JS. 2004. Adipose tissue as an endocrine organ. J Clin Endocrinol Metab 89: 2548–2556.

Kho EM, McCowan LME, North RA, Roberts CT, Chan E, Black MA, Taylor RS, Dekker GA. 2009. Duration of sexual relationship and its effect on preeclampsia and small for gestational age perinatal outcome. J Reprod Immunol 82: 66–73.

Kim J-M, Hong K, Lee JH, Lee S, Chang N. 2009. Effect of folate deficiency on placental DNA methylation in hyperhomocysteinemic rats. J Nutr Biochem 20(3): 172–176.

Kim SY, Dietz PM, England L, et al. 2007. Trends in pre-pregnancy obesity in nine states, 1993–2003. Obesity 15: 986–993.

Klisch K, Boos A, Friedrich M, Herzog M, Feldmann M, Sousa NM, Beckers JF, Leiser R, Schuler G. 2006. The glycosylation of pregnancy-associated glycoproteins and prolactin-related protein-I in bovine binucleate trophoblast giant cells changes before parturition. Reprod 132: 791–798.

Knipp GT, Liu B, Audus KL, Fujii H, Ono T, Soares MJ. 2000. Fatty acid transport regulatory proteins in the developing rat placenta and in trophoblast cell culture models. Placenta 21: 367–375.

Knox K, Baker JC. 2008. Genomic evolution of the placenta using co-option and duplication and divergence. Genome Res 18: 695–705.

Knudsen A, Lebech M, Hansen M. 1995. Upper gastrointestinal symptoms in the third trimester of the normal pregnancy. Eur J Obstet Gyn Reprod Biol 60(1): 29–33.

Koelman CA, Coumans ABC, Nijman HW, Doxiadis IIN, Dekker GA, Claas FHJ. 2000. Correlation between oral sex and a low incidence of preeclampsia: a role for soluble HLA in seminal fluid? J Reprod Med 46: 155–166.

Kohler PF, Farr RS. 1966. Elevation of cord over maternal IgG immunoglobulin: evidence for an active placental IgG transport. Nature 210: 1070–1071.

Kojima M, Hosoda H, Date Y, Nakazato M, Matsuo H, Kangawa K. 1999. Ghrelin is a growth-hormone releasing acylated peptide from stomach. Nature 402: 656–660.

Korebrits C, Ramirez MM, Watson L, Brinkman E, Bocking AD, Challis JRG. 1998. Maternal corticotrophin-releasing hormone is increased with impending preterm birth. J Clin Endocrinol Metab 83(5): 1585–1591.

Kristal MB, Thompson AC, Grishkat HL. 1985. Placenta ingestion enhances opiate analgesia in rats. Physiol Behav 35(4): 481–486.

Kupfer A, Müller H, Antoniazzi MM, Jared C, Greven H, Nussbaum RA, Wilkinson M. 2006. Parental investment by skin feeding in a caecilian amphibian. Nature 440: 926–929.

Kuzawa CW. 1998. Adipose tissue in human infancy and childhood: an evolutionary perspective. Am J Phys Anthropol 107(27): 177–209.

Laatikainen T, Virtanen T, Kaaja R, Salminen-Lappalainen K. 1991. Corticotropin-releasing hormone in maternal and cord plasma in pre-eclampsia. Eur J Obstet Gyn Reprod Biol 39(1): 19–24.

Lander ES, Linton LM, Birren B, Nusbaum C, Zody MC, et al. 2001. Initial sequencing and analysis of the human genome. Nature 409: 860–921.

Larqué E, Krauss-Eischmann S, Campoy C, Hartl D, Linde J, Klingler M, Demmelmair H, Caño A, Gil A, Bondy B, Koletzko B. 2006. Docosahexaenoic acid supply in pregnancy affects placental expression of fatty acid transport proteins. Am J Clin Nutr 84: 853–861.

Lash GE, Robson SC, Bulmer JN. 2010. Review: Functional role of uterine natural killer (uNK) cells in human early pregnancy decidua. Placenta 24 Suppl A: S87–S92.

Lauder JM, Sze PY, Krebs H. 1981. Maternal influences on tryptophan hydroxylase activity in embryonic rat brain. Dev Neurosci 4: 291–295.

Laurin M, Reisz RR, Girondot M. 2000. Caecilian viviparity and amniote origins: a reply to Wilkinson and Nussbaum. J Nat Hist 34(2): 311–315.

Leddy MA, Power ML, Schulkin J. 2008. Maternal obesity and the impact on maternal and fetal health. Rev Obstet Gynecol 1: 170–178.

Leiser R, Dantzer V. 1988. Structural and functional aspects of porcine placental microvasculature. Anat Embryol 177: 409–419.

Leonard WR, Snodgrass JJ, Robertson ML. 2007. Effects of brain evolution on human nutrition and metabolism. Ann Rev Nutr 27: 311–327.

Leung TN, Chung TKH, Madsen G, Lam PKW, Sahota D, Smith R. 2001. Rate of rise in maternal plasma corticotrophin-releasing hormone and its relation to gestational length. Brit J Obstet Gynaec 108: 527–532.

Lewis K, Li C, Perrin MH, Blount A, Kunitake K, Donaldson C, Vaughan TM, Reyes TM, Gulyas J, Fischer W, et al. 2001. Identification of urocortin III, an additional member of the corticotropin-releasing factor (CRF) family with high affinity for the CRF2 receptor. PNAS 98: 7570–7575.

Li E, Bestor TH, Jaenisch R. 1992. Targeted mutation of the DNA methyltransferase gene results in embryonic lethality. Cell 69: 915–926.

Little J. 2009. Invited commentary: Maternal effects in preterm birth—effects of maternal genotype, mitochondrial DNA, imprinting, or environment? Am J Epidemiol 170: 1382–1385.

Lockwood CJ. 2004. The initiation of parturition at term. Obstet Gynecol Clin North Am 31(4): 935–947.

Lombardi J, Wourms JP. 1985. The trophotaenial placenta of a viviparous goodeid fish. II. Ultrastructure of trophotaeniae, the embryonic component. J Morphol 184: 293–309.

Long JA. 2011. Dawn of the deed. Sci Am 304: 34–39.

Long JA, Trinajstic K, Young GC, Senden T. 2008. Live birth in the Devonian period. Nature 453: 650–652.

Longo LD, Reynolds LP. 2010. Some historical aspects of understanding placental development, structure and function. Int J Dev Biol 54: 237–255.

Lovejoy DA, Balment RJ. 1999. Evolution and physiology of the corticotropin-releasing factor (CRF) family of neuropeptides in vertebrates. Gen Comp Endocrinol. 115: 1–22.

Lucchesi JC, Kelly WG, Panning B. 2005. Chromatin remodeling in dosage compensation. Genetics 39: 615–651.

Lynch CM, Sexton DJ, Hession M, Morrison JJ. 2008. Obesity and mode of delivery in primigravid and multigravid women. Am J Perinat 25: 163–167.

Lyon MF. 1961. Gene action in the X-chromosome of the mouse (Mus musculus L). Nature 190: 372–373.

Macklon NS, Geraedts JPM, Fauser BCJM. 2002. Conception to ongoing pregnancy: the "black box" of early pregnancy loss. Hum Reprod Update 8: 333–343.

Makrigiannakis A, Margioris AN, Le Goascogne C, Zoumakis E, Nikas G, Stoumaras C, Psychoyos A, Gravanis A. 1995. Cotricotropin-releasing hormone (CRH) is expressed at the implantation sites of early pregnant rat uterus. Life Sci 57(20): 1869–1875.

Martin JA, Kirmeyer S, Osterman M, Shepherd RA. 2009. Born a bit too early: recent trends in late preterm birth. NCHS Data Brief 24: 1–8.

Martin RD. 2003. Human reproduction: a comparative background for medical hypotheses. J Reprod Immunol 59: 111–135.

Mastorakos G, Scopa CD, Vryonidou A, Friedman TC, Kattis D, Phenekos C, Merino MJ, Chrousos GP. 1994. Presence of immunoreactive corticotrophin-releasing hormone in normal and polycystic human ovaries. J Clin Endocrinol Metab 79(4): 1191–1197.

Maston GA, Ruvolo M. 2002. Chorionic gonadotropin has a recent origin within primates and an evolutionary history of selection. Mol Biol Evol 19(3): 320–335.

Matsubara K, Tarui H, Toriba M, Yamada K, Nishida-Umehara C, Agat K, Matsuda Y. 2006. Evidence for different origin of sex chromosomes in snakes, birds, and mammals and step-wise differentiation of snake sex chromosomes. PNAS 103(48): 18190–18195.

Mazzotta P, Stewart DE, Koren G, Magee LA. 2001. Factors associated with elective termination of pregnancy among Canadian and American women with nausea and vomiting of pregnancy. J Psychosom Obst Gyn 22: 7–12.

McClintock B. 1950. The origin and behavior of mutable loci in maize. PNAS 36: 344–355.

McEwen BS. 1998. Stress, adaptation, and disease: allostasis and allostatic load. Ann NY Acad Sci 840: 33–44.

McEwen BS. 2004. Protection and damage from acute and chronic stress: allostasis and allostatic overload and relevance to the pathophysiology of psychiatric disorders. Ann NY Acad Sci 1032: 1–7.

McIntyre HD, Serek R, Crane DI, Veveris-Lowe T, Parry A, Johnson S, Leung KC, Ho KKY, Bougoussa M, Hennen G, et al. 2000. Placental growth hormone (GH), GH-binding protein, and insulin-like growth factor axis in normal, growth-retarded, and diabetic pregnancies: correlations with fetal growth. J Clin Endocrinol Metab 85(3) 1143–1150.

McLean M, Bisits A, Davies J, Woods R, Lowry P, Smith R. 1995. A placental clock controlling the length of human pregnancy. Nat Med 1: 460–463.

Medawar PB. 1953. Some immunological and endocrinological problems raised by the evolution of viviparity in vertebrates. Symp Soc Exp Biol 7: 320–338.

Mess A, Carter AM. 2007. Evolution of the placenta during early radiation of placental mammals. Comp Biochem Phys A 148: 769–779.

Messerli M, May K, Hansson SR, Schneider H, Holzgreve W, Hahn S, Rusterholtz C. 2010. Feto-maternal interactions in pregnancies: placental microparticles activate peripheral blood monocytes. Placenta 31(2): 106–112.

Mikkelsen TS, Wakefield MJ, Aken B, Amemiya CT, Chang JL, et al. 2007. Genome of the marsupial Monodelphis domestica reveals innovation in non-coding sequences. Nature 447: 167–177.

Miller-Lindholm AK, LaBenz CJ, Ramey J, Bedows E, Ruddon RW. 1997. Human chorionic gonadotropin-ß gene expression in first trimester placenta. Endocrinology 138: 5459–5465.

Milligan LA. 2005. Concentration of sIgA in the milk of *Macaca mulatta*. Am J Phys Anthropol 128: 153.

Mills JL, Troendle J, Conley MR, Carter T, Druschel CM. 2010. Maternal obesity and congenital heart defects: a population-based study. Am J Clin Nutr 91: 1543–1549.

Minas et al. 2007. Abortion is associated with increased expression of FasL in deciduate leukocytes and apoptosis of extravillous trophoblasts: a role for CRH and urocortin. Mol Hum Reprod 13: 663–673.

Mincheva-Nilsson L, Baranov V. 2010. The role of placental exosomes in reproduction. Am J Reprod Immunol 63: 520–533.

Mishima T, Kurasawa G, Ishikawa G, Mori M, Kawahigashi Y, Ishikawa T, Luo S-S, Takizawa T, Goto T, Matsubara S, et al. 2007. Endothelial expression of Fc gamma receptor llb in the full-term human placenta. Placenta 28: 170–174.

Mittal P, Hassan SS, Espinoza J, Kusanovic JP, Edwin S, Gotsch F, Erez O, Than N, Mazaki-Tovi S, Romero R. 2008. The effect of gestational age and labor on placental growth hormone in amniotic fluid. Growth Horm IGF Res 18(2): 174.179.

Mittwoch U. 1978. Parthenogenesis. J Med Genet 15: 165–181.

Mold JE, Venkatasubrahmanyam S, Burt TD, Michaëlsson J, Rivera JM, Galkina SA, Weinberg K, Stoddart CA, McCune JM. 2010. Fetal and adult hematopoietic stem cells give rise to distinct T cell lineages in humans. Science 330(6011): 1695–1699.

Moore T, Haig D. 1991. Genomic imprinting in mammalian development: a parental tug-of-war. Trends Genet 7: 45–49.

Mossman HW. 1926. The rabbit placenta and the problem of placental tissue. Am J Anat 37(3): 433–497.

Mossman HW. 1937. Comparative morphogenesis of the fetal membranes and accessory uterine structures. Contrib Embryol (Carnegie Institute) 26: 129–246.

Mossman HW. 1987. Vertebrate fetal membranes. New Brunswick, NJ: Rutgers University Press; Basingstoke, UK: Macmillan Press

Muglia LJ. 2000. Genetic analysis of fetal development and parturition control in the mouse. Pediatr Res 47: 437–443.

Muglia LJ, Katz MK. 2010. The enigma of spontaneous preterm birth. N Eng J Med 362: 529–535.

Munkhaugen J, Vikse BE. 2009. New aspects of pre-eclampsia: lessons for the nephrologist. Nephrol Dial Transplant 24: 2964–2967.

Muramatsu Y, Sugino N, Suzuki T, Totsune K, Takahasi K, Tashiro A, Hongo M, Oki Y, Sasano H. 2001. Urocortin and corticotropin-releasing factor receptor expression in normal cycling human ovaries. J Clin Endocrinol Metab 86(3): 1362–1369.

Naeye R. 1987. Do placental weights have clinical significance? Hum Pathol 18: 387–391.

Nakamura M. 2009. Sex determination in amphibians. Semin Cell Dev Biol 20(3): 271–282.

Namekawa SH, Lee JT. 2009. XY and ZW: Is meiotic sex chromosome inactivation the rule in evolution? PLoS Genet 5(5): 1–3.

Ng HK, Novakovic B, Hiendleder S, Craig JM, Roberts CT, Saffery R. 2010. Distinct patterns of gene-specific methylation in mammalian placentas: implications for placental evolution and function. Placenta 31(4): 259–268.

Njogu A, Owiti GO, Persson E, Oduor-Okelo D. 2006. Ultrastructure of the chorioallantoic placenta and chorionic vesicles of the lesser bush baby (*Galago senegalensis*). Placenta 27(6): 771–779.

Numan M, Fleming AS, Levy F. 2006. Maternal behavior. In Knobil and Neill's Physiology of Reproduction, ed. JD Neill, 1921–1993. 3rd ed. St. Louis: Elsevier.

O'Brien B, Naber S. 1992. Nausea and vomiting during pregnancy: effects on the quality of women's lives. Birth 19(3): 138–143.

Oftedal OT. 2002a. The mammary gland and its origin during synapsid evolution. J Mammary Gland Biol 7: 225–252.

Oftedal OT. 2002b. The origin of lactation as a water source for parchment-shelled eggs. J Mammary Gland Biol 7: 253–266.

Oftedal OT, Bowen WD, Boness DJ. 1993. Energy transfer by lactating hooded seals and nutrient deposition in their pups during the four days from birth to weaning. Physiol Zool 66: 412–436.

Ogata M, Hasegawa Y, Ohtani H, Mineyama M, Miura I. 2008. The ZZ/ZW sex-determining mechanism originated twice and independently during evolution of the frog, *Rana rugosa*. Heredity 100(1): 92–99.

Ohno S. 1967. Sex chromosomes and sex-linked genes. Springer-Verlag, Berlin and New York.

Ohno S, Kaplan WD, Kinosita R. 1959. Formation of the sex chromatin by a single X-chromosome in liver cells of *Rattus norvegicus*. Exp Cell Res 18: 415–418.

Okamoto I, Heard E. 2009. Lessons from comparative analysis of X-chromosome inactivation in mammals. Chromosome Res 17: 659–669.

Padykula HA, Taylor JM. 1977. Uniqueness of the bandicoot chorioallantoic placenta (Marsupialia: Peramelidae). Cytological and evolutionary interpretations. In Reproduction and evolution, ed. JH Calaby, CH Tyndale-Biscoe, 303–313. Canberra: Australian Academy of Science.

Palomar L, DeFranco EA, Lee KA, Allsworth JE, Muglia LJ. 2007. Paternal race is a risk factor for preterm birth. Am J Obstet Gynecol 197: 152.e1–152.e7.

Papper A, Jameson NM, Romero R, Weckle AL, Mittal P, Benirschke K, Santolaya-Forgas J, Uddin M, Haig D, Goodman M, et al. 2009. Ancient origin of placental expression in the growth hormone genes of anthropoid primates. PNAS 106(40): 17083–17088.

Partridge C, Shardo J, Boettcher A. 2007. Osmoregulatory role of the brood pouch in the euryhaline Gulf pipefish, *Syngnathus scovelli*. Comp Biochem Phys A 147(2): 556–561.

Pellicer J, Fay MF, Leitch IJ. 2010. The largest eukaryotic genome of them all? Botanical J Linnean Soc 164: 10–15.

Pencharz RI, Long JA. 1931. The effect of hypophysectomy on gestation in the rat. Science 74(1912): 206.

Pepper G, Roberts SC. 2006. Rates of nausea and vomiting in pregnancy and dietary characteristics across populations. Proc R Soc B 273: 2675–2679.

Perkins AV, Eben F, Wolfe CDA, Schulte HM, Linton EA. 1993. Plasma measurements of corticotrophin-releasing hormone-binding protein in normal and abnormal human pregnancy. J Endocrinol 138: 149–157.

Petraglia F, Florio P, Vale WW. 2005. Placental expression of neurohormones and other neuroactive molecules in human pregnancy. In Birth, Distress and Disease, ML Power and J Schulkin (eds), 16–73.

Pijnenborg R, Vercruysse L. 2004. Thomas Huxley and the rat placenta in the early debates on evolution. Placenta 25: 233–237.

Pijnenborg R, Vercruysse L. 2006. Mathias Duval on placental development in mice and rats. Placenta 27: 109–118.

Pijnenborg R, Vercruysse L. 2007. Erasmus Darwin's enlightened views on placental fusion. Placenta 28: 775–778.

Pijnenborg R, Vercruysse L, Carter AM. 2011. Deep trophoblast invasion and spiral artery remodelling in the placental bed of the chimpanzee. Placenta 32: 400–408.

Plaks V, Brinberg T, Berkutzki T, Sels S, Yashar AB, Kalchenko V, Mor G, Keshet E, Dekel N, Neeman M, et al. 2008. Uterine DCs are crucial for decidua formation during embryo implantation in mice. J Clin Invest 118: 3954–3965.

Policastro PF, Daniels-McQueen S, Carle G, Boime I. 1986. A map of the hCG beta–LH beta gene cluster. J Biol Chem 261: 5907–5916.

Ponce de León MS, Golovanova L, Doronichev V, Romanova G, Akazawa T, Kondo O, Ishida H, Zollikofer CPE. 2008. Neanderthal brain size at birth provides insights into the evolution of human life history. PNAS 105(37): 13764–13768.

Power ML, Bowman ME, Smith R, Zieglar TE, Layne DG, Schulkin J, Tardiff SD. 2006. The pattern of maternal serum corticotropin-releasing hormone concentration during pregnancy in the common marmoset (*Callithrix jacchus*). Am J Primatol 68: 181–188.

Power ML, Schulkin J. 2005. Introduction: brain and placenta, birth and behavior, health and disease. In Birth, Distress and Disease, ed. ML Power, J Schulkin, 1–15. Cambridge, UK: Cambridge University Press.

Power ML, Schulkin J. 2006. Functions of corticotropin-releasing hormone in anthropoid primates: from brain to placenta. Am J Hum Biol 18(4): 431–447.

Power ML, Schulkin J. 2008. Sex differences in fat storage, fat metabolism, and the health risks from obesity: possible evolutionary origins. Br J Nutr 99: 931–940.

Power ML, Schulkin J. 2009. The Evolution of Obesity. Baltimore: Johns Hopkins University Press.

Power ML, Williams LE, Gibson SV, Schulkin J, Helfers J, Zorilla EP. 2010. Pattern of maternal circulating CRH in laboratory-housed squirrel and owl monkeys. Am J Primatol 72: 1004–1012.

Profet M. 1992. Pregnancy sickness as adaptation: a deterrent to maternal ingestion of teratogens. In The Adapted Mind, ed. JH Barkow, L Cosmides, J Tooby, 327–366. Oxford: Oxford University Press.

Profet M. 1993. Menstruation as a defense against pathogens transported by sperm. Q Rev Biol 68: 335–386.

Rainey WE, Rehman KS, Carr BR. 2004a. Fetal and maternal adrenals in human pregnancy. Obstet Gynecol Clin North Am 31: 817–835.

Rainey WE, Rehman KS, Carr BR. 2004b. The human fetal adrenal: making adrenal androgens for placental estrogens. Semin Reprod Med 22(4): 327–336.

Ramírez-Pinilla MP, De Pérez G, Carreño-Excobar JF. 2006. Allantoplacental ultrastructure of an Andean population of *Mabuya* (Squamata, Scincidae). J Morphol 267: 1227–1247.

Ramsey EM, Corner GW Jr, Donner MW, Stran HM. 1960. Radioangiographic studies of circulation in the maternal placenta of the rhesus monkey: preliminary report. PNAS 46(7): 1003–1008.

Rawn SM, Cross JC. 2008. The evolution, regulation, and function of placenta-specific genes. Annu Rev Cell Dev Biol 24: 159–181.

Reece EA. 2008. Obesity, diabetes, and links to congenital defects: a review of the evidence and recommendations for intervention. J Matern Fetal Neonatal Med 21: 173–180.

Reik W, Lewis A. 2005. Co-evolution of X-chromosome inactivation and imprinting in mammals. Nat Rev Genet 6: 403–410.

Reik W, Walter J. 2001. Genomic imprinting: parental influence on the genome. Nat Rev Genet 2: 21–32.

Reilly SM, White TD. 2003. Hypaxial motor patterns and the function of epipubic bones in primitive mammals. Science 299(5605): 400–402.

Renfree MB. 2000. Maternal recognition of pregnancy in marsupials. Rev Reprod 5: 6–11.

Renfree MB, Ager EI, Shaw G, Pask AJ. 2008. Genomic imprinting in marsupial placentation. Reprod 136: 523–531.

Renfree MB, Shaw G. 1996. Reproduction of a marsupial: from uterus to pouch. Anim Reprod Sci 42(1): 393–403.

Rens W, O'Brien PCM, Grützner F, Clarke O, Graphodatskaya D, Tsend-Ayush E, Trifonov VA, Skelton H, Wallis MC, Johnston S, et al. 2007. The multiple sex chromosomes of platypus and echidna are not completely identical and several share homology with the avian Z. Genome Biol 8: R243.

Rich TH, Vickers-Rich P, Trusler P, Flannery TF, Cifelli R, Constantine A, Kool L, Van Klaveren N. 2001. Monotreme nature of the Australian Early Cretaceous mammal *Teinolophos*. Acta Palaeontol Pol 46(1): 113–118.

Robbins JR, Skrzypczynska KM, Zeldovich VB, Kapidzic M, Bakardjiev AI. 2010. Placental syncytiotrophoblast constitutes a major barrier to vertical transmission of *Listeria monocytogenes*. PLoS Pathog 6(1): 1–13.

Robert KA, Thompson MB, Seebacher F. 2003. Facultative sex allocation in the viviparous lizard *Eulamprus tympanum*, a species with temperature-dependent sex determination. Aust J Zool 51: 367–370.

Roberts CT. 2010. IFPA award in placentology lecture: complicated interactions between genes and the environment in placentation, pregnancy outcome and long term health. Placenta 24: S47–S53.

Roberts JM, Redman CWG. 1993. Pre-eclampsia: more than pregnancy-induced hypertension. Lancet 341: 1447–1451.

Robillard P-Y, Hulsey TC, Dekker GA,Chaouat G. 2003. Preeclampsia and human reproduction: an essay of long term reflection. J Reprod Immunol 59: 93–100.

Robinson BG, Emanuel RL, Frim DM, Majzoub JA. 1988. Glucocorticoid stimulates expression of corticotrophin-releasing hormone gene in human placenta. PNAS 85(14): 5244–5248.

Rockwell C, Vargas E, Moore LG. 2003. Human physiological adaptation to pregnancy: inter- and intraspecific persectives. Am J Hum Bio 15: 330–341.

Roseboom TJ, Painter RC, de Rooij SR, van Abeelen AFM, Veenendaal MVE, Osmond C, Barker DJP. 2011. Effects of famine on placental size and efficiency. Placenta 32: 395–399.

Ross MT, Grafham DV, Coffey AJ, Scherer S, McLay K, Muzny D, Platzer M, Howell GR, Burrows C, Bird CP, et al. 2005. The DNA sequence of the human X chromosome. Nature 434 (7031): 325–337.

Ruth V, Hallman M, Laatikainen T. 1993. Corticotropin-releasing hormone and cortisol in cord plasma in relation to gestational age, labor, and fetal distress. Am J Perinatol 10(2): 115–118.

Rutherford JN, Eklund A, Tardif SD. 2009. Placental insulin-like growth factor II (IGF-II) and its relation to litter size in the common marmoset monkey (*Callithrix jacchus*). Am J Primatol 71: 969–975.

Rutherford J, Tardif S. 2008. Placental efficiency and intrauterine resource allocation strategies in the common marmoset pregnancy. Am J Phys Anthropol 137: 60–68.

Rutherford JN, Tardif SD. 2009. Developmental plasticity of the microscopic placenta architecture in relation to litter size variation in the common marmoset monkey (*Callithrix jacchus).* Placenta 30: 105–110.

Sacks G, Sargent I, Redman C. 1999. An innate view of human pregnancy. Immunol Today 20(3): 114–118.

Sagebakken G, Ahnesjö I, Mobley KB, Gonçalves IB, Kvarnemo C. 2010. Brooding fathers, not siblings, take up nutrients from embryos. Proc R Soc B 277(1683): 971–977.

Sakamoto T, McCormick SD. 2006. Prolactin and growth hormone in fish osmoregulation. Gen Comp Endocrinol 147: 24–30.

Samuel CA, Perry JS. 1972. The ultrastructure of pig trophoblast transplanted to an ectopic site in the uterine wall. J Anat 113: 139–149.

Sargent IL, Borzychowski AM, Redman CWG. 2006. NK cells and human pregnancy—an inflammatory view. Trends Immunol 27(9): 1471–4906.

Sarre SD, Georges A, Quinn A. 2004. The ends of a continuum: genetic and temperature-dependent sex determination in reptiles. BioEssays 26: 639–645.

Sasaki A, Liotta AS, Luckey MM, Margioris AN, Suda T, Krieger DT. 1984. Immunoreactive corticotropin-releasing factor is present in human maternal plasma during the third trimester of pregnancy. J Clin Endocrinol Metab 59: 812–814.

Sasaki Y, Ladner DG, Cole LA. 2008. Hyperglycosylated human chorionic gonadotropin and the source of pregnancy failures. Fertil Steril 89(6): 1781–1786.

Scammell JG, Funkhouser JD, Moyer FS, Gibson SV, Willis DL. 2008. Molecular cloning of pituitary glycoprotein alpha-subunit and follicle stimulating hormone and chorionic gonadotropin beta-subunits from new world squirrel monkey and owl monkey. Gen Comp Endocrinol 155(3): 534–541.

Schulkin J. 1999. Corticotropin-releasing hormone signals adversity in both the placenta and the brain: regulation by glucocorticoids and allostatic overload. J Endocrinol 161: 349–356.

Seal US, Sinha AA, Doe RP. 1972. Placental iron transfer: relationship to placental anatomy and phylogeny of the mammals. Am J Anat 134: 263–269.

Seasholtz AF, Valverde RA, Denver RJ. 2002. Corticotropin-releasing hormone-binding protein: biochemistry and function from fishes to mammals. J Endocrinol 175: 89–97.

Segre LS, O'Hara MW, Arndt S, Stuart S. 2007. The prevalence of postpartum depression: the relative significance of three social status indices. Soc Psych Psych Epid 42(4): 316–321.

Seligman CG, Murray MA. 1911. Note upon an early Egyptian standard. Man 2: 165–171.

Selye H, Collip JB, Thomson DL. 1933. On the effect of the anterior pituitary-like hormone on the ovary of the hypophysectomized rat. Endocrinology 17(5): 494–500.

Seres J, Bornstein SR, Seres P, Willenberg HS, Schulte KM, Scherbaum WA, Ehrhart-Bornstein M. 2004. Corticotropin-releasing hormone system in human adipose tissue. J Clin Endocrinol Metab 89: 965–970.

Šerman L, Vlahović M, Šijan M, Bulić-Jakuš F, Šerman A, Sinčić N, Matijević R, Jurić-Lekić G, Katušić A. 2007. The impact of 5-azacytidine on placental weight, glycoprotein pattern and proliferating cell nuclear antigen expression in rat placenta. Placenta 28(8): 803–811.

Sherman GB, Wolfe MW, Farmerie TA, Clay CM, Threadgill DS, Sharp DC, Nilson JH. 1992. A single gene encodes the beta-subunits of equine luteinizing hormone and chorionic gonadotropin. Mol Endocrinol 6(6): 951–959.

Sherman PW, Flaxman SM. 2002. Nausea and vomiting of pregnancy in an evolutionary perspective. Am J Obstet Gynecol 186: S190–S197.

Shine R, Elphick MJ, Donnellan S. 2002. Co-occurrence of multiple, supposedly incompatible modes of sex determination in a lizard population. Ecol Lett 5(4): 486–489.

Sinclair EA, Pramuk JB, Bezy RL, Crandall KA, Sites JW Jr. 2010. DNA evidence for non-hybrid origins of parthenogenesis in natural populations of vertebrates. Evolution 64: 1346–1357.

Sifakis S, Papadopoulou E, Konstantinidou A, Giahnakis E, Fragouli Y, Karkavitsas N, Koumantakis E, Kalmanti M. 2009. Increased levels of human placental growth hormone in the amniotic fluid of pregnancies affected by Down syndrome. Growth Horm IGF Res 19: 121–125.

Simpson JL. 2007. Causes of fetal wastage. Clin Obstet Gynecol 50: 10–30.

Sirianni R, Mayhew BA, Carr BR, Parker CR Jr, Rainey WE. 2005. Corticotropin-releasing hormone (CRH) and urocortin act through type 1 CRH receptors to stimulate dehydroepiandrosterone sulfate production in human fetal adrenal cells. J Clin Endocrinol Metab 90: 5393–5400.

Slominski A, Wortsman J. 2000. Neuroendocrinology of the skin. J Clin Endocrinol Metab 21(5): 457–487.

Smith JDL, Gregory TR. 2009. The genome sizes of megabats (Chiroptera: Pteropodidae) are remarkably constrained. Biol Lett 5: 347–351.

Smith JJ, Voss SR. 2007. Bird and mammal sex-chromosome orthologs map to the same autosomal region in a salamander (*Ambystoma*). Genetics 177: 607–613.

Smith R, Chan E-C, Bowman ME, Harewood WJ, Phippard AF. 1993. Corticotropin-releasing hormone in baboon pregnancy. J Clin Endocrinol Metab 76: 1063–1068.

Smith R, Mesiano S, Chan E-C, Brown S, Jaffe RB. 1998. Corticotropin-releasing hormone directly and preferentially stimulates dehydroepiandrosterone sulfate secretion by human fetal adrenal cortical cells. J Clin Endocrinol Metab 83(8): 2916–2920.

Smith R, Mesiano S, Nicholson R, Clifton V, Zakar T. 2005. The regulation of human parturition. In Birth, distress and disease: placenta-brain interactions, ed. ML Power, J Schulkin, 74–87. Cambridge, UK: Cambridge University Press.

Smith R, Wickings EJ, Bowman ME, Belleoud A, Dubreuil G, Davies JJ, Madsen G. 1999. Corticotropin-releasing hormone in chimpanzee and gorilla pregnancies. J Clin Endocrinol Metab 84(8): 2820–2825.

Smith SR, de Jonge L, Pellymounter M, Nguyen T, Harris R, York D, Redmann S, Rood J, Bray GA. 2001. Peripheral administration of human corticotropin-releasing hormone: a novel method to increase energy expenditure and fat oxidation in man. J Clin Endocrinol Metab 86: 1991–1998.

Soares MJ, Müller H, Orwig KE, Peters TJ, Dai G. 1998. The uteroplacental prolactin family and pregnancy. Biol Reprod 58: 273–284.

Sockman KW, Schwabl H. 2000. Yolk androgens reduce offspring survival. Proc R Soc B 267: 1452–1456.

Speake BK, Herbert JF, Thompson MB. 2004. Evidence for placental transfer of lipids during gestation in the viviparous lizard, *Pseudemoia entrecasteauxii*. Comp Biochem Phys A 139(2): 213–220.

Spong CY, Iams J, Goldenberg R, Hauck FR, Willinger M. 2011. Disparities in perinatal medicine: preterm birth, still bith, and infant mortality. Obstet Gynecol 117(4): 948–955.

Springer MS, Murphy WJ, Eizirik E, O'Brien S. 2005. Molecular evidence for major placental clades. In The Rise of Placental Mammals: Origins and Relationships of Major Extant Clades, ed. KD Rose, JD Archibald, 37–49.

Stothard KJ, Tennant PWG, Bell R, Rankin J. 2009. Maternal overweight and obesity and the risk of congenital anomalies. JAMA 301(6): 636–650.

Stoye JP. 2006. Koala retrovirus: a genome invasion in real time. Genome Biology 7: 241.

Strahl H. 1899. Der Uterus gravidus von Galago agisymbanus. Abh Senckenb Naturf Ges 26: 155–199.

Strahl H. 1905. Beiträge zur vergleichenden Anatomie der Placenta. Abh Senckenb Naturf Ges 27: 263–319.

Sugimoto J, Schust DJ. 2009. Human endogenous retroviruses and the placenta. Reprod Sci 16(11): 1023–1033.

Sullivan JL. 1981. Iron and the sex difference in heart disease risk. Lancet 1: 1293–1294.

Swanson LD, Bewtra C. 2008. Increase in normal placental weights related to increase in maternal body mass index. J Matern Fetal Neonatal Med 21: 111–113.

Tafuri A, Alferink J, Moller P, Hammerling GJ, Arnold B. 1995. T-cell awareness of paternal alloantigens during pregnancy. Science 270: 630–633.

Takagi N, Sasaki M. 1975. Preferential inactivation of the paternally derived X chromosome in the extraembryonic membranes of the mouse. Nature 256: 640–642.

Tardif SD. 1997. The bioenergetics of parental behavior and the evolution of alloparental care in marmosets and tamarins. In Cooperative Breeding in Mammals, ed. NG Solomon, JA French. Cambridge, UK: Cambridge University Press.

Tardif SD, Ziegler TE, Power M, Layne DG. 2005. Endocrine changes in full-term pregnancies and pregnancy loss due to energy restriction in the common marmoset (*Callithrix jacchus*). J Clin Endocrinol Metab 90: 335–339.

Tarlington RE, Meers J, Young PR. 2006. Retroviral invasion of the koala genome. Nature 442: 79–81.

Tita ATN, Rouse DJ. 2010. Progesterone for preterm birth prevention: an evolving intervention. Am J Obstet Gynecol 202(6): e10–1.

Trivers RL. 1972. Parental investment and sexual selection. In Sexual Selection and the Descent of Man, ed. B Campbell, 136–179. Chicago: Aldine.

Trivers RL. 1974. Parent-offspring conflict. Am Zool 14(1): 249–264.

Tyson E. 1698. The anatomy of an opossum. Phil Trans. RS London. J Theor Biol 139(3): 343–357.

Üstün Ç. 2003. Dr. Thomas Wharton's eponyms: Wharton's duct and Wharton's jelly. Turk J Med Ethics 11(3).

Vale W, Spiess J, Rivier J. 1981. Characterization of a 41-residue ovine hypothalamic peptide that stimulates secretion of corticotrophin and β-endorphin. Science 213: 1394–1397.

Vanneste E, Voet T, Le Caignec C, Ampe M, Konings P, Melotte C, Debrock S, Amyere M, Vikkula M, Schuit F, et al. 2009. Chromosome instability is common in human cleavage-stage embryos. Nat Med 15: 577–583.

Verberg MFG, Gillott DJ, Al-Fardan N, Grudzinskas JG. 2005. Hyperemesis gravidarum, a literature review. Hum Reprod Update 11: 527–539.

Veyrunes F, Waters PD, Miethke P, Rens W, McMillan D, Alsop AE, Grützner F, Deakin JE, Whittington CM, Schatzkamer K, et al. 2008. Bird-like sex chromosomes of platypus imply recent origin of mammal sex chromosomes. Genome Res 18: 965–973.

Vickers MH, Breier BH, Cutfield WS, Hofman PL, Gluckman PD. 2000. Fetal origins of hyperphagia, obesity, and hypertension and postnatal amplification by hypercaloric nutrition. Am J Physiol Endocrinol Metab 279: E83–E87.

Vikanes A, Skjaerven R, Grjibovski AM, Gunnes N, Vangen S, Magnus P. 2010. Recurrence of hyperemesis gravidarum across generations: population based cohort study. BMJ 340: c2050.

Vlahovic M, Bulić-Jakuš F, Jurić-Lekić G, Fučić A, Marić S, Šerman D. 1999. Changes in the placenta and in the rat embryo caused by the demethylating agent 5-azacytidine. Int J Dev Biol 43: 843–846.

von Haeckel E. 1866. Generelle Morphologie der Organismen: allgemeine Grundzüge der organischen Formen-Wissenschaft, mechanisch begründet durch die von C. Darwin reformirte Decendenz-Theorie. Berlin: Verlag von Georg Reimer.

Waddington CH. 1942. Canalization of development and the inheritance of acquired characters. Nature 150: 563–565.

Wake MH, Dickie R. 1998. Oviduct structure and function and reproductive modes in amphibians. J Exper Zool 282: 477–506.

Wake N, Takagi N, Sasaki M. 1976. Non-random inactivation of X chromosome in the rat yolk sac. Nature 262: 580–581.

Wallis OC, Wallis M. 2006. Evolution of growth hormone in primates: the GH gene clusters of the new world monkeys marmoset (*Callithrix jacchus*) and white-fronted capuchin (*Cebus albifrons*). J Mol Evol 63(5): 591–601.

Wang CN, Chang SD, Peng HH, Lee YS, Chang YL, Cheng PJ, Chao AS, Wang TH, Wang

HS. 2010. Change in amniotic fluid levels of multiple anti-angiogenic proteins before development of preeclampsia and intrauterine growth restriction. J Clin Endocrinol Metab 95(3): 1431–1441.

Warren JE, Silver RM. 2008. Genetics of pregnancy loss. Clin Obstet Gynecol 51(1): 84–95.

Warren WB, Patrick SL, Goland RS. 1992. Elevated maternal plasma corticotropin-releasing hormone levels in pregnancies complicated by preterm labor. Am J Obstet Gynecol 166: 1198–1204.

Waterland RA, Jirtle RL. 2003. Transposable elements: targets for early nutritional effects on epigenetic gene regulation. Mol Cell Biol 23(15): 5293–5300.

Weigle RM, Weigle MM. 1989. Nausea and vomiting of early pregnancy and pregnancy outcome. A meta-analytic review. BJOG—Int J Obstet Gy 96: 1312–1318.

West JD, Frels WI, Chapman VM, Papaioannou VE. 1977. Preferential expression of the maternally derived X chromosome in the mouse yolk sac. Cell 12: 873–882.

White TD. 1989. An analysis of epipubic bone function in mammals using scaling theory. J Theor Biol 139(3): 343–357.

Wilcox AJ, Skjærven R, Terje Lie, R. 2008. Familial patterns of preterm delivery: maternal and fetal contributions. Am J Epidemiol 167: 474–479.

Wildman DE. 2011. Review: toward an integrated evolutionary understanding of the mammalian placenta. Placenta 32, Suppl. 2: S142–S145.

Wildman DE, Chen C, Erez O, Grossman LI, Goodman M, Romero R. 2006. Evolution of the mammalian placenta revealed by phylogenetic analysis. PNAS 103: 3203–3208.

Wilkins JF, Haig D. 2003. What good is genomic imprinting: the function of parent-specific gene expression. Nat Rev Genet 4: 359–368.

Wilkins JM, Hill S. 2006. Food in the Ancient World. Malden, MA: Blackwell Publishing.

Wilkins-Haug L. 2008. Assisted reproductive technology, congenital malformations, and epigenetic disease. Clin Obstet Gynecol 51(1): 96–105.

Wimsatt WA, Enders AC, Mossman HW. 1973. A reexamination of the chorioallantoic placenta membrane of a shrew, *Blarina brevicauda*: resolution of a controversy. Am J Anat 138(2): 207–233.

Wooding FBP. 1984. Role of binucleate cells in fetomaternal cell fusion at implantation in the sheep. Am J Anat 170(2): 233–250.

Wooding FB. 1992. Observations on the morphogenesis, cytochemistry, and significance of the binucleate giant cells of the placenta of ruminants. Am J Anat 89: 233–281.

Wooding FBP, Ramirez-Pinilla MP, Forhead AS. 2010. Functional studies of the placenta of the lizard *Mabuya* sp. (Scincidae) using immunocytochemistry. Placenta 31: 675–685.

Wooding P, Burton G. 2008. Comparative placentation: structures, functions and evolutions. Berlin: Springer.

Wourms JP. 1981. Viviparity: the maternal-fetal relationship in fishes. Am Zool 21(2): 473–515.

Wu WX, Unno S, Giussani DA, Mecenas CA, McDonald TJ, Nathanielsz PW. 1995. Corticotropin-releasing hormone and its receptor distribution in fetal membranes and placenta of the rhesus monkey in late gestation and labor. Endocrinology 136: 4621–4628.

Yang R, You X, Tang X, Gao L, Ni X. 2006. Corticotropin-releasing hormone inhibits progesterone production in cultured human placental trophoblasts. J Mol Endocrinol 37: 533–540.

Yavarone MS, Shuey DL, Sadler TW, Lauder JM. 1993. Serotonin uptake in the ectoplacental cone and placenta of the mouse. Placenta 14: 149–161.

Ye C, Li Y, Shi P, Zhang Y. 2005. Molecular evolution of growth hormone gene family in old world monkeys and hominoids. Gene 250(2): 183–192.

Yen SSC. 1994. The placenta as the third brain. J Reprod Med 39: 277–80.

York TP, Straus JF III, Neale MC, Eaves LJ. 2010. Racial differences in genetic and environmental risk to preterm birth. PLoS ONE 5: e12391.

Zhang Y, Cantor RM, MacGibbon K, Romero R, Goodwin TM, Mullin PM, Fejzo MS. 2011. Familial aggregation of hyperemesis gravidarum. Am J Obstet Gynceol 204(3): 230.e1–7.

Zinaman MJ, Clegg ED, Brown CC, O'Connor J, Selevan SG. 1996. Estimates of human fertility and pregnancy loss. Fertil Steril 65(3): 503–309.

Zoumakis E, Chatzaki E, Charalampopoulos I, Margioris AN, Angelakis E, Koumantakis E, Gravanis A. 2001. Cycle and age-related changes in corticotropin-releasing hormone levels in human endometrium and ovaries. Gynecol Endocrinol 15(2): 98–102.

Zoumakis E, Margioris AN, Stournaras C, Dermitzaki E, Angelakis E, Makrigiannakis A, Koumantakis E, Gravanis A. 2000. Corticotrophin-releasing hormone (CRH) interacts with inflammatory prostaglandins and interleukins and affects the decidualization of human endometrial stroma. Mol Hum Reprod 6(4): 344–351.

INDEX

aardvarks, 13, 16, 72, 84, 88, 94
abortion, 22, 88, 178, 182, 187, 189, 191, 202–5, 206, 207, 215. *See also* pregnancy loss, early
adaptations, 2, 3, 14; early pregnancy loss, 203–4, 210; egg retention, 51, 56–57; imprinted genes, 19; placental structural variations, 17; to produce live-born young, 12; related to bipedalism, 96; in response to pathogens, 10–11; selective forces created by, 7; specifically human, 17; to support brain growth, 10, 17; X chromosome inactivation, 8, 18
adipocytes, 186, 188
adiponectin, 178, 188
adipose tissue, 12, 106, 110; CRH and, 175, 178; cytokines produced by, 188–89; as endocrine organ, 106, 186–87, 225; gestational weight gain and, 184, 190; leptin and, 170–72; obesity and, 23, 187, 188; placenta and, 188–89
adrenocorticotropic hormone (ACTH), 172, 174, 179, 180
Afrotheria, 71, 72, 84, 86, 94
aldosterone, 179–80
allantois, 12, 15, 19, 29, 53–54, 55, 74, 76, 82, 151, 154
allostasis, 164
allostatic overload, 186
altricial offspring, 73, 224
amnion, 12, 15, 19, 29, 53, 55, 74, 76, 151, 154
amniotes, 12, 15, 53–56, 223; birds, 59–60; gene imprinting in, 132–33; sex chromosomes of, 122–28; synapsids and sauropsids, 63–66; VT genes in, 70
amniotic egg, 2, 12, 15, 19, 53–56, 57, 63, 223
amniotic fluid, 29, 35, 38, 48, 53; GH2 in, 168; leptin in, 171–72
amniotic sac, 15, 29, 53, 102, 154, 214, 223
amphibians, 3, 53, 54, 63, 70, 124, 175; eggs of, 2, 52, 53, 63; external fertilization in, 52; live birth of, 51, 58–59; sex chromosomes of, 123, 124, 125
anapsids, 64
ancestral placenta, 14, 15, 72, 81, 91–93, 97
anencephalic fetus, 180
Angelman syndrome, 153
angiosperms, 133
anteaters, 16, 72, 84, 93, 94
anthropoid primates, 90, 203; brain of, 10, 108, 118, 228; CG in, 21, 74, 158–62, 172, 173, 202; clade of, 96; CRH in, 176–82; decidualization in, 90; DHEAS in, 114, 179–80; GH genes in, 156–58, 168; HPA axis in, 179–80; IgG transfer to fetus, 17, 115–17; invasive placenta in, 110, 211, 213, 228; menstruation of, 211–12; neonates of, 73; placental lactogens in, 156–57; placental signaling in, 21, 113–14; placental type in, 3, 14, 16, 17, 40, 78, 86, 89, 93, 96–100, 109, 113, 117; pregnancy loss in, 202; prolactin gene in, 157; retroviral genes in, 81, 148; superficial implantation in, 228; syncytiotrophoblast in, 78, 81, 101
anthropoid stage of primate placentation, 97, 98
antigens: cancer-testis, 124; fetal, 22, 86, 87, 193; human leukocyte, 194, 220; maternal, 115, 193; maternal immune response to, 88, 193, 195, 212; MHC, 87, 192; paternal, 87, 115, 130, 193, 220; uterine exposure to, 211–12
apes, 97, 107, 197, 198, 199, 203, 219; bipedalism in, 109, 110, 118; brain of, 110, 118, 228; CG genes in, 160; clade of, 96, 98, 203; CRH in, 21, 177, 180–81; implantation in, 3, 40, 98, 102, 109, 110, 117, 213, 216, 228–29; placental type in, 3, 14, 17

arachidonic acid (AA), 107–8
Aranzi, Giulio Cesare, 36
areolae, 86
Aristotle, 11, 25, 31–35, 61, 210
armadillos, 43, 70, 72, 93, 94; placental type in, 15–16, 43, 84, 89, 93, 94
artiodactyls. *See* cetartiodactyls
assisted reproductive technology (ART), 152–53
autoimmune diseases, 195
autosomes, 122–25, 132

baboons, 43, 96, 113, 168, 190; CRH in, 176, 177, 180–82, 229; leptin in, 170–71
Baker, Julie, 154–55
bandicoots, 82
Barker, D. J. P., 185
Barr, Murray, 126
bats, 72, 90, 94, 143, 170; placental type in, 16, 85, 88, 89, 92, 113, 224
bears, 14, 16, 72, 224
Beckwith-Wiedemann syndrome, 153
Benirschke, Kurt, 43, 83
Bernard, Claude, 45, 163–64
Bertram, Ewart George, 126
Beutler, Ernest, 126
biological anthropology, 8, 10
bipedalism, 90, 96, 107, 109–10, 117–18, 220
birds, 3, 13, 59–60, 120, 133; eggs of, 2, 12, 15, 53, 55, 60, 62, 69; evolution of, 63, 64, 123; genomes of, 143, 147; sex chromosomes of, 18, 120, 122–28; VT genes in, 70
birth weight, 168–69, 171, 186, 205
blastocyst, 40; with chromosomal abnormalities, 203–4; decidualization and, 90, 211, 213; human, 99, 102, 103, 213; implantation of, 3, 192, 201, 204, 209, 210, 212, 213; invasion of uterine epithelium by, 3, 77, 82–83, 213–15; male reproductive strategy and, 209–10; maternal immune system and, 46, 212, 216, 226; primate, 97–98, 198; TGCs and, 77
blastomere, 154
blastula, 154
Boreoeutheria, 71
Boulton, Matthew, 39
brain, 9, 20, 45, 224; chimpanzee, 10, 108; dolphin calf, 108, 227; fetal, 20, 106–8, 172, 183, 227; gene expression and development of, 131, 133, 140; glucose requirement of, 112, 170, 227; guinea pig, 227–28; horse, 42; human vs. ape, 96, 108, 110, 113, 228; maternal, 167, 170; mosaic evolution and, 90; placenta-brain axes, 20; preeclampsia and, 220; serotonin in, 183; sex determination and, 120, 124
brain growth, 10, 107–8; expensive tissue hypothesis of, 107, 227; fatty acids and, 107–8; placental invasiveness and, 110–11; placental type and, 10, 17, 109, 112–13, 116–18, 213, 227–28, 229; postnatal, 108
Brown-Sequard, C. E., 45
bushbabies, 40, 100, 168–69

caecilians, 58
Caenorhabditis elegans, 143, 146, 175
Cannon, Walter, 164
cardiovascular disease, 22–23, 53, 185, 211, 221
carnivores, 64, 73, 107, 175; placental type in, 14, 16, 40, 78, 88–89, 94, 99
cats, 16, 72, 126
cauls, 29–31
cell differentiation, 153–54
cercopithicines, 96
cetaceans, 16
cetartiodactyls, 14, 16, 17, 72, 78, 86, 88, 94, 97, 98
chickens, 12, 69, 123, 124
child rearing, 200
chimpanzees, 10, 90, 108, 198, 203; CG in, 160, 161; CRH in, 176, 177, 179, 180; IgG transfer in, 116, 117; implantation in, 117, 228; interbirth interval in, 199; menstruation in, 212; placenta of, 109, 110, 228–29; preeclampsia in, 220; pregnancy loss in, 202
chorio-allantoic placentas, 12, 14, 15, 62, 70, 71, 76, 82, 86, 223, 224
chorion, 12, 15, 19, 29, 53–54, 55, 57, 74, 76, 82–83, 151, 154
chorionic gonadotropin (CG), 74, 114, 158–62, 172–73; equine, 74, 87–88, 159, 160, 161; genes for, 160, 161, 173; hemochorial placenta and, 161; stimulation of ovarian progesterone production by, 21, 74–75; structure of, 159–60, 172. *See also* human chorionic gonadotropin
chorionic sac, 102, 106
chorionic villi, 15, 102, 103, 104, 106, 172
chorion laeve, 106

chorio-vitteline placentas, 15, 69, 82, 223
chromosomes: abnormalities of, 202–5; autosomes, 132; crossover of, 138; sex, 7–8, 18, 120, 122–31
circulation, maternal and fetal, 5, 74, 106; cell layers between, 75, 81–83, 86, 103, 113, 114; equine CG and, 88, 161; fatty acid transfer between, 108; glucose transfer between, 105; historical concepts of, 32, 36, 37–38; interhemal distance and, 74–75; leptin and, 171; regulatory molecules and, 113–14, 163; serotonin and, 184; in sharks, 61; syncytiotrophoblast and, 79–80; TGCs and, 77
Civinni, Fillipo, 37
coelacanths, 12, 53, 57–58
colobines, 96, 181
colugos, 85, 94, 98, 161
Columbus, Renaldus, 34, 35
corpus luteum, 21, 156, 173, 204–5
corticotropin-releasing hormone (CRH), 21, 158–59, 169, 172, 173–82, 217, 219; in anthropoid primate placenta, 177–80; CRH peptide family, 173, 175; evolution of, 175; as neuropeptide, 173–75; patterns of placental expression of, 80, 180–82, 229; in peripheral tissues, 175, 177; placental, 176–80; postpartum level of, 184; preterm birth and, 21, 178, 217, 219; receptors for, 173–74, 175, 176, 177, 219; as reproductive hormone, 175–76
corticotropin-releasing hormone binding protein (CRH-BP), 173, 175, 177
cortisol, 172, 177–79, 186, 187, 217; adrenal production of, 174, 179–80; in human pregnancy, 164, 177, 180
cotyledonary placental type, 14–16, 34, 84, 94
cows, 77, 99, 130, 137, 156–57, 158; placental type in, 16, 34, 84, 88, 91, 98
C-reactive protein, 190
Creighton, Harriet, 138
crocodilians, 13, 55, 64, 66, 120, 123
cynodonts, 66, 68
cytogenetics, 152
cytokines, 5, 78, 102, 166, 225; adipose tissue secretion of, 186–89; placental production of, 5, 20, 25, 45, 77, 104, 142; pro-inflammatory, 189–90, 194, 217
cytomegalovirus, 215
cytotrophoblasts, 76, 77, 78–81; CG expression by, 173; extravillous, 76, 102, 104–5, 110, 114, 172–73, 184, 213–16, 225, 228; fusion of, 148, 172; villous, 172–73

Darwin, Charles, 38, 39
Darwin, Erasmus, 38, 39
da Vinci, Leonardo, 33, 34, 36
de Bordeu, Theophile, 45
decidua, 15, 90–91, 103, 104; communication between blastocyst and, 204, 211; cytotrophoblast invasion of, 172, 173, 176, 215; deep invasion of, 117, 221; definition of, 90, 192, 201; engulfment of blastocyst by, 213, 214; formation of, 90–91, 102; shallow invasion of, 101, 113; TGC-stimulated formation of, 77; uterine NK cells of, 194–95, 220
decidua basalis, 102, 103, 104, 106
decidua capsularis, 40, 102, 103, 106, 214
decidualization, 77, 83, 90–91, 109, 176, 192–93, 212
decidua parietalis, 103
deer, 14, 16, 25, 48
dehydroepiandrosterone sulfate (DHEAS), 114, 179–80
depression, postpartum, 47–48, 184
diabetes mellitus: birth outcomes and, 112, 170; GH and, 168; in-utero programming of, 53, 185; leptin and, 170; obesity and, 187, 190
diapsids, 64
diffuse placental type, 15–16; endotheliochorial, 92; epitheliochorial, 14, 16, 42, 85–86, 92, 93, 94, 97, 100
dinosaurs, 13, 55, 63, 64, 66, 72
Diogenes of Apollonia, 31
discoid placental type, 15–16, 94; hemochorial, 14, 84–85, 92, 94, 97–98, 99–100, 109; hemomonochorial, 100–102; human, 102, 106, 109
DNA, 12, 25, 119, 230; ancient retroviral, 10, 147–49, 152; epigenetics and, 131–32, 190–91; genetic regulation and, 19, 140–45, 149–52, 165–66, 191, 226–27; junk, 144; mammalian, 143; mitochondrial, 134; mobile (transposon), 138, 145–47; paternal, 134; phenotype and, 9; of placental cells, 142; pregnancy loss and, 202–3; repeated sequences of, 142; of sex chromosomes, 125–26; of TGCs, 77; transcription

DNA *(cont.)*
of, 79, 144–45, 150, 152, 166; X chromosome inactivation and, 20, 125, 132. *See also* genome
DNA methylation, 19, 131, 132, 141, 142, 145, 151, 152, 153, 191; adaptive function of, 147; IVF and, 153; in placental tissues, 148, 151
docosahexaenoic acid (DHA), 107–8
dogs, 44, 63, 70, 72, 157; placental type in, 16, 34, 36, 40, 84
dolphins, 32, 108, 224, 227; placental type in, 14, 16–17, 85–86, 98, 108, 109
Down syndrome, 168, 204
Duval, Matthias, 39

echidnas, 62, 67, 69, 122
eclampsia, 219
Edentata, 72
eggs, 2, 12; amniotic, 2, 12, 15, 19, 53–56, 57, 63, 223; amphibian, 2, 52, 53, 63; bird, 2, 12, 15, 53, 55, 60, 62, 69; crocodilian, 55; deposited in environment, 51; external fertilization of, 52–53, 58; fish, 2, 52, 53; internal fertilization of, 13, 52, 55, 57, 58, 60; marsupial, 82, 223; maternal immune response to, 56–57; monotreme, 62, 69, 82, 223; platypus, 62, 69; reptile, 2, 12, 53, 55; retention of, 51, 55, 56–57, 60, 62, 69
Egypt, ancient, 25, 28–29
electron microscopy, 11, 25, 41, 43, 44, 88
elephants, 16, 72, 78, 84, 88–89, 94
embryonic period, 202
embryophagy, 60
Enders, Allen, 43, 214
endocrine organ: adipose tissue as, 186–87; placenta as, 9, 12, 20–21, 44–46, 48
endometrial cups, 87–88, 99
endotheliochorial placentas, 14, 15, 16, 40, 41, 83–85, 88–89, 97, 98, 99, 111; hemophagous zones of, 89, 111; immune function of, 115; phylogeny and, 92–94; signaling of, 114; syncytiotrophoblast in, 78, 81, 89; transformation to hemochorial placenta, 89, 92–93, 99, 113, 224
epigenetics, 131–32, 155, 224, 226; in-utero programming of disease and, 190–91; IVF and, 152–53; mechanisms of, 152, 191
epipubic bones, 68
epitheliochorial placentas, 14–16, 17, 40–42, 78, 81, 82–88, 90; dolphin, 108, 109, 227; galago, 100, 116; horse, 42, 87–88, 112, 159; immunologic advantage of, 98–99, 115; lemur and loris, 94, 97, 114, 116; nutritional efficiency of, 86, 111; phylogeny and, 91, 92–94, 97, 98, 109; pig, 75, 83, 86; signaling by, 114; species with, 84–86, 91, 98
equine chorionic gonadotropin (eCG), 74, 87–88, 159, 160, 161
estrogens, 21, 45, 114, 179, 180, 184
Euarchontoglires, 71, 72
eutherian mammals, 2, 3, 13, 14–16, 68–69. *See also* placental mammals
evolution, convergent, 6, 17, 92, 99, 159, 160
evolutionary biology of placentas, 2–3, 4–6, 13, 96–118, 223–24; expressed genes, 155–62; Grosser's view, 41–42
exocoelemic cavity, 102, 104
extinct species, 13, 59, 61, 64–66, 68, 71–72, 107
extraembryonic fetal membranes, 1, 12, 15, 19, 29, 53, 54, 55, 76, 141, 148, 151, 154; of birds, 59; gene expression and, 155; of mammals, 63, 74; of monotremes, 82

Fabricius ab Aquapendente, 34, 36
Fallopius, Gabriele, 34, 36
fat, 186; deposits of, in body, 186; dietary, 186, 198; gestational weight gain as, 190; leptin and, 170; metabolism of, 172, 175; placental development and, 106–8; visceral, 190. *See also* adipose tissue
fathers, 1, 134–36, 200, 210; genetic contribution of, 22, 191; influence of, on preterm birth, 217–18; parent-offspring conflict, 3, 9, 19, 23, 74, 119, 133–37, 169, 210, 214, 227; paternal X chromosome inactivation, 8, 18, 128, 129–30, 132, 137; reproductive fitness of, 52, 134, 137, 213
fatty acids, 105, 107–8, 190
Fc receptors, 117
felines. *See* cats
fenestrae, temporal, 63, 64
ferrets, 16, 40
fertility, 197–98
fertilization: in amphibians, 58; in angiosperms, 133; in cephalopods, 31; decidualization and, 90–91, 212; external, 52–53, 58; implantation and, 3, 170, 200–204, 212–13; internal, 13, 52, 55, 57, 58, 60; in-vitro, 152–53, 203; male

reproductive strategy and, 209; in mammals, 154; pregnancy loss and, 200–202; in reptiles, 59; in sharks, 60–61
fetal period, 202
fetal wastage, 201
fishes, 158, 165, 198, 207; bony, 57–58; cartilaginous, 60; coelacanth, 12, 53, 57–58; CRH in, 173, 175; eggs of, 2, 52, 53; genomes of, 147; GH/prolactin and osmoregulation in, 156, 158–59; live birth of, 12, 13, 50–51, 56, 57, 64; omega-3 fatty acids in, 108; placoderms, 12–13, 50, 64; sex reversals in, 120; teleost, 158
Flexner, Louis, 43
follicle-stimulating hormone (FSH), 160, 172
Forsdahl, A., 185
fossils, 4, 65; of amniotic eggs, 55–56; cetartiodactyl, 16; coelacanth, 12, 53, 57; cynodont, 66; live birth and, 50, 51, 68; monotreme, 67; placoderm, 12, 50; primate, 87; retroviral remnants as, 147, 149
Frazer, James George, 28, 29
frogs, 53, 58–59, 123, 143
fruit flies, 127, 128, 143

galagos, 17, 100, 168–69; placental type in, 85, 100, 116
Galen, 32, 34–36, 38, 210–11
gene expression, 25, 119, 149; in different organs, 9, 46, 166; epigenetics and, 131, 152; gene imprinting and, 132; IVF and, 153; jumping genes and, 139; monoallelic, 132; nutritional factors and, 145; obesity and, 190–91; placental, 15, 46, 80, 97, 141–42, 151–52, 154–55; placental-specific gene function and, 155–56; regulation of, 144–45, 149–52, 226–27; of retroviral genes, 10, 17, 20, 93, 142, 147–50, 152; of sex chromosomes, 18–19, 127, 130–31, 144; in syncytiotrophoblast nuclei, 80; of TGCs, 76; X chromosome, 18–19, 127, 130–31, 144
gene imprinting, 3, 19, 22, 132–33, 191, 213; adaptive function of, 19, 46, 119, 137, 226–27; in angiosperms, 133; IVF and, 153; in marsupials, 132, 137, 226; parthenogenesis and, 133; trophoblast, 194; X chromosome inactivation and, 8, 137
genetic regulation, 3, 8, 10, 15, 19–20, 149–52, 166, 226–27; cytogenetics, 152; epigenetic, 131–32, 152–53, 191; gene imprinting, 132–33; McClintock's studies of, 139, 138–41, 146. *See also* gene imprinting
genome, 131–32, 152–53, 226; epigenetics and, 131–32, 152–53, 226; gene imprinting and, 19, 132–33, 137; jumping genes and, 138, 145–47; junk DNA in, 144; maize, 138–41; parent-offspring conflict, 19, 119, 134–35, 227; placenta and, 2, 5, 9, 19, 25, 92, 93, 97, 134, 141–42, 151, 153; preterm birth and, 217; regulation of, 144–45, 153; retroviruses in, 10, 19, 81, 147–50; sequencing of, 91, 143, 152, 165–66, 226; size of, 143–44, 152; VT genes in, 70. *See also* DNA
genomics, 24, 25, 46
gestational challenges, 6, 7, 14, 23, 94–95, 196–222; early pregnancy loss, 200–205; fertility, efficiency, and child rearing, 197–200; invasion of uterine tissue by human placenta, 213–15; morning sickness, 205–10; placental infections, 215–16; preeclampsia, 219–22; preterm birth, 216–19
ghrelin, 167
gibbons, 40, 177, 181, 228, 229
glucocorticoids, 174, 178, 179–80, 187
glucose, 9, 20, 59, 169, 225; brain requirement for, 112, 170, 227; GH2 and, 167–69, 170; maternal levels of, 164, 169–70, 177, 190; placental regulation of transfer of, 105, 112; poor maternal control of, 186, 187; teratogenicity of, 105, 112. *See also* diabetes mellitus
glucose-6-phosphate dehydrogenase deficiency, 126
gorillas, 87, 109, 110, 198, 220, 228, 229; CRH in, 176, 177, 179, 180, 181
Graves' disease, 195
Greece, ancient, 25, 31–32
Grosser, Otto, 40–42, 44, 83
growth hormone (GH), 21, 113, 156–58, 184; evolution of, 155–56, 169; gene duplications for, 156–57, 168; glucose and, 167–69, 170; IUGR and, 168–69; leptin and, 171; in osmoregulation, 156, 158–59; pituitary (GH1), 157, 166–67, 168; placental (GH2), 157, 166–70; primate genes for, 21, 93, 114, 155–56, 168–69; receptor for, 167; in rodents, 167; in sheep, 168
guinea pigs, 85, 110, 111, 177, 227–28

Haig, David, 137, 227
Harvey, William, 36–37, 44
hectocotyl arm of cephalopods, 31
hedgehogs, 39, 72
HELLP syndrome, 220
hemochorial placentas, 14, 15, 17, 83, 86, 89–90, 91, 95, 97, 228; adaptive value of, 111; in anthropoid primates, 97–98, 99–100, 109, 113, 117, 161; in armadillos, 43, 93, 94; brain size and, 227; CG and, 161; Grosser's classification, 40, 41, 42; IgG transfer by, 115–17; nutritional efficiency of, 111–13; signaling by, 114; species with, 84–85, 89, 91, 93, 94, 98, 99–100; syncytiotrophoblast in, 78, 81, 89; transformation of endotheliochorial placenta to, 89, 92–93, 99, 113, 224
hemophagous zones, 40, 89, 111, 113
hemotrophic nutrition, 86, 104, 106, 223
heterogametic sex, 18, 121, 122, 123, 124, 125, 126, 128
Hill, J. P., 97–98
hippopotami, 16
histiotrophic nutrition, 86, 104, 106, 223
histone modifications, 131, 132, 141, 145, 152, 153, 191
historical investigations, 11–12, 24–49
homeostasis, 164–65
homogametic sex, 18, 121, 124
horses, 42, 72, 73; CG in, 74, 87–88, 159, 160, 161; chorionic girdle in, 87; CRH in, 177; endometrial cups of, 87–88, 99; placental products from, 25–27, 48; placental type in, 16, 17, 42, 83, 86, 87–88, 91, 94, 97, 98, 112, 114, 159, 227
HPA axis, fetal, 179–80
Hubrecht, Ambrosius, 39
human chorionic gonadotropin (hCG), 19, 21, 160–62, 169, 172–73, 204; evolution of, 161, 173; hyperglycosylated, 161, 173, 204, 208; levels in pregnancy, 80, 172; morning sickness and, 21, 173, 208–9; in successful gestation, 161–62, 173; trisomy 21 and, 204–5
human leukocyte antigen (HLA) system, 194, 220
human placenta, 1, 4, 17, 102–6, 227–30; ancient retroviral genes and, 142, 148; bipedalism and, 109–10, 117–18; CG production by, 160, 173; CRH expression by, 80, 176; evolution of, 2, 7, 8, 14, 17, 96–118, 228; fatty acid transport by, 108; Fc receptors of, 117; fetal brain growth and, 10, 108, 113, 117–18, 228, 229; gene expression by, 155; growth and development of, 104–5; historical investigations of, 11, 28, 34–36, 40; IgG transfer by, 115–17, 229; infections of, 215–16; interhemal distance in, 75–76; invasive qualities of, 87, 110–11, 213–15, 221; nutritional efficiency of, 111–13; pathogens and, 11, 17, 116; progesterone production by, 204; serotonin and, 183–84; signaling by, 20–21, 113–14; syncytins and, 148; syncytiotrophoblast of, 17, 76–81, 172; terms for, 34; TGCs in, 77; type of, 10, 75, 102, 109; uniqueness of, 4, 10, 96, 224; X chromosome inactivation in, 130
Hunter, William and John, 32, 37, 38
Huxley, Thomas, 38–39
hyenas, 84, 89, 94, 99
hyperemesis gravidarum, 206
hypertension, 6, 196; in-utero programming of, 53, 185; obesity and, 187; preeclampsia, 6, 7, 23, 105, 196, 219–22, 229
hypoxia, 105

imaging technology, 11, 225
immune function molecules, 5, 9, 20, 22, 45, 166
immune system: fetal, 22, 193; immune cells in adipose tissue, 186; of seahorse, 52
immune system, maternal, 191–95; CG and, 173; in decidualization, 192; of horse, 87; immunological advantages of pregnancy, 195; inflammatory response of, 189; passage of disease history to fetus, 17, 115; placental immunology, 22; placental infections and, 215–16; placental protection of fetus from, 5, 12, 22, 46, 57, 78, 98, 99, 192–93; preeclampsia and, 22, 189, 191, 194, 220, 222; regulatory function of, 8, 9, 22, 166; tolerance of implanted embryo, 21, 173, 193–95; uterine NK cells, 192–95, 212, 216, 220, 221, 225
immunoglobulin (IgG) transfer to fetus, 17, 82, 115–16, 223, 229
implantation, 2, 17, 154, 192–93, 196, 198; aggressive, 213–15; CG and, 75, 161–62, 172, 202, 204, 205, 209; CRH and, 175–76, 229; decidualization and, 90, 103, 106, 212; endotheliochorial placentation and, 88; epitheliochorial placentation and, 83, 86; failure of, 200–202, 204, 211, 212; in humans, 102–6, 109, 161, 166, 192–93, 218, 227–29; as inflammatory state, 193, 213;

interstitial, 3, 18, 40, 98, 109, 110, 113, 114, 117, 161–62, 213–14, 216, 220, 227–29; leptin and, 170; male reproductive strategy and, 209–10; maternal-fetal conflict and, 134; maternal immune system and, 9, 21, 22, 46, 99, 192, 193, 220, 225; menstruation and, 212, 216, 222; morning sickness and, 210; in primates, 43, 97, 101, 110, 113, 114, 117, 202, 211, 219, 228–29; progesterone and, 204, 209; and protection against pathogens, 18, 229; in rodents, 76–77; superficial, 228–29; TGCs in, 76–77; trisomy 21 and, 204–5; uterine epithelial preparation for, 192–93, 201, 212; zona pellucida and, 154
infections, placental, 215–16
inflammation: CRH and, 175–76; inflammatory cytokines, 189–90, 217; obesity and, 188, 189–90; preeclampsia and, 191; pregnancy as state of, 46, 175–76, 189, 193, 213
information molecules, 5, 12, 45, 102, 166–84; CG, 172–73; CRH, 173–82; leptin, 170–72; placental GH, 166–70; placental production of, 5, 9, 12, 20, 22, 45, 78, 88, 142, 156, 163, 166, 225; placentophagy and, 184; secretion of, by decidua, 90, 211; serotonin, 183–84; TGCs and, 77
insectivores, 39, 66, 67, 86, 89, 98
insects, 63, 65, 143, 175
insulin, 9, 45, 164, 178, 184, 186; CRH and, 177–78; leptin and, 171
insulin resistance, 164–65, 167, 169, 170, 190
interbirth intervals, 199, 218, 220
interhemal distance, 74–75, 81, 83, 113
intrauterine growth restriction (IUGR), 7, 104–5, 215, 218; CRH and, 178; GH and, 168–69; preeclampsia and, 171, 218, 221, 222, 226
in-utero origins of disease, 22–23, 185–86
in-vitro fertilization (IVF), 152–53, 203
iron, 27, 89, 111, 112–13, 211

Jirtle, Randy, 145
jumping genes, 138–40, 142, 145–47

kangaroos, 14
Keys, Ancel, 158
killer cell immunoglobulin-like receptors (KIRs), 220
Klebs, Edwin, 40
Knox, Kirstin, 154–55
koala retrovirus (KoRV), 150
Krogh, August, 158

lactation, 69, 165, 184; brain growth during, 10; in mammals, 62, 69, 70, 74, 137, 165, 198, 223–24; in marsupials, 62, 82; preceding live birth, 70; preceding placentation, 69; in primates, 198; weaning and, 10, 198, 199, 200
lagomorphs, 72, 81, 82, 94, 98
large-for-gestational-age fetuses, 171, 186
Laurasiatheria, 71, 72, 86
Lavoisier, Antoine, 37, 38
lemuroid stage of primate placentation, 97
lemurs, 72, 94, 98, 160, 176–77; placental type in, 3, 16, 17, 40, 85, 86, 87, 93–94, 97, 114, 116
leptin, 169, 170–72, 184; fetal, 171; maternal levels of, 164, 171, 184, 190; obesity and, 188, 190; placental production of, 170–71; receptors for, 171, 190; in sheep, 170, 172
light microscopy, 11, 41, 42–43
Linnaeus, 131
Listeria monocytogenes, 18, 215–16
live birth, 50–71, 223; of amphibians, 58–59; amniotic eggs and, 53–56; of bony fishes, 57–58; egg retention and, 56–57; evolutionary history of, 12–13, 50–51, 223; external fertilization and, 52–53; lactation and, 69; of mammals, 61–69; maternal mortality and, 196; placental immunology and, 22, 98, 191; of reptiles, 59; sex determination and, 121–22; of sharks, 50, 60–61; species with, 13, 51; trisomy 21 rate for, 204
lizards, 13, 55, 59, 64, 66, 133; sex determination in, 18, 120, 121, 122, 123
llamas, 72
long interspersed nuclear elements (LINE-1s), 124–25, 129, 144
lorises, 98, 160, 177, 181; placental type in, 3, 16, 17, 85, 86, 93–94, 97, 114, 116
Lunar Society, 39
luteinizing hormone (LH), 159–61, 172
Lyon, Mary Francis, 126

maize genome, 138–41
major histocompatibility complex (MHC), 87, 192
mammals: comparative placentation in, 72, 73–95; embryo retention in, 69–71; evolutionary

mammals *(cont.)*
history of, 63–69; live birth of, 61–63; molecular phylogeny of, 71–72; neonates of, 73; placental, 68–69; teeth of, 63
marmosets, 207, 212; CG in, 160; CRH in, 176, 177, 180–82; GH genes in, 157; placental type in, 43, 85, 100–101
marsupials, 13, 62, 67, 71; eggs of, 82, 223; embryonic cells of, 130, 154; epipubic bones of, 68; gene imprinting in, 132, 137, 226; live birth of, 62; placenta of, 14–15, 62–63, 69, 82; sex chromosomes of, 122, 124, 128–29; VT genes in, 70; X chromosome inactivation in, 8, 18, 128, 129, 130, 137
maternal-fetal conflict, 19, 23, 133–37, 169
maternal-placental-fetal unit, 1–3, 133–35
maternal survival, 165
Mayow, John, 37
McClintock, Barbara, 138–41, 146
Medawar, Peter, 46, 192
meiosis, 18, 122, 124, 125–26, 132, 203
menstruation, 35, 91, 210–13; pathogen-cleansing function of, 211, 216
metabolic syndrome, 178, 186
metatherian mammals, 13, 133. *See also* marsupials
mice, 145, 193; CRH knockout, 176; decidualization in, 192–93; genome of, 143, 144; placental gene expression in, 154–55; placental type in, 85, 89; prolactin gene in, 156–57; serotonin in, 183–84; syncytins in, 79; X chromosome inactivation in, 126, 129, 137
microtome, 40
mineralocorticoids, 179–80
Minot, Charles, 40
moles, 72, 86, 87
monkeys: capuchin, 157; CG in, 160; clades of, 96; colobus, 96, 181; CRH in, 21, 176–77, 180–81, 229; dusky leaf, 160; Galen's dissection of, 32; GH genes in, 157, 168; marmoset, 43, 85, 100–101, 157, 160, 176, 177, 180–82, 207, 212; menstruation in, 212; New World (platyrrhine), 3, 96, 98, 100, 118, 160, 177, 181; Old World (catarrhine), 3, 96, 98, 100, 160, 177, 181; owl, 160, 176, 177, 180; placental type in, 3, 14, 17, 98; Rhesus, 42, 43, 85, 101–2, 116, 117, 157, 176, 177; spider, 157, 160, 168; squirrel, 115–16, 177, 180; vervet, 212
monosomy X, 202
monotremes, 14, 62, 67–68, 132, 137, 226; eggs of, 62, 69, 82, 223; epipubic bones of, 68; placenta of, 62, 69, 82; sex chromosomes of, 122, 123, 128
morning sickness, 6, 7, 23, 205–10, 229; adaptive value of, 7, 209–10; hCG and, 21, 173, 208–9; as protective against pregnancy loss, 23, 206, 207
morula, 154
mosaic evolution, 89–90, 91
Mossman, Harland, 42
multiple sclerosis, 195

natural killer (NK) cells, 191; uterine, 192–94, 212, 216, 220, 221, 225
natural selection, 211, 230
nematodes, 143
Noortwyck, Wilhelm, 37
nutrient transfer to embryo, 9, 38, 98, 224; in amphibians, 58; gene expression and, 145; hemotrophic, 86, 104, 106, 223; histiotrophic, 86, 104–6, 223; in humans, 104–6; leptin and, 170; in live-born invertebrates, 50; matrotrophic, 82, 223; placental regulation of, 12, 25, 31, 36–37, 40, 44; in reptiles, 59; in sharks, 61; "thrifty phenotype hypothesis," 185–86; VT genes and, 70
nutritional efficiency, 111–13
nutritional supplements, placental, 25–27, 47–49
nutrition of offspring after birth: lactation and, 62, 69, 74, 82, 198–99, 223–24; in primates, 198; weaning and, 10, 198, 199, 200. *See also* lactation

obesity, 23, 186–88, 190; epigenetics and, 190; fetal programming of, 185–86; as inflammatory state, 188, 189–90; leptin receptors in, 171, 190; during pregnancy, 107, 187–90
Ohno, Susumo, 126
oophagy, 60
orangutans, 40, 117, 198, 228, 229
osmoregulation in fish, 156, 158–59
ovoviviparity, 50–51, 57, 59
ovulation, 197

oxidative stress, 104, 106, 112
oxygen, 9, 20, 42, 44, 65, 75, 106, 225; for brain growth, 112, 227; delivery of, to fetus, 37, 38, 74, 105–6, 111, 112; discovery of, 36, 37, 38

parasites, 10, 17, 116
parental investment, 136
parent-offspring conflict, 3, 9, 19, 23, 74, 119, 133–37, 169, 210, 214, 227
parthenogenesis, 22, 133
pathogens, 10–11, 20, 116, 193, 211–14, 226; interstitial implantation and, 18, 229; menstruation and, 211, 216; placental infection by, 215–16; sperm-borne, 122, 216; transmission of maternal exposure history to fetus, 17–18
pelycosaurs, 64
perissodactyls, 17, 72, 94, 97, 98
phenotype, 9, 20, 81, 145, 209; epigenetics and, 131; placental, 118, 161; pregnancy metabolic, 164; sex, 120–22, 125–26; "thrifty phenotype hypothesis," 185–86; of uterine NK cell expression, 194
phylogeny, 2, 6, 14–15, 38–39, 41, 86, 91, 98, 123–24; of CRH receptors, 174; fertility and, 197; molecular, of mammals, 71–72; placenta and, 88, 92–95, 109, 117; retrotransposons and, 149; of vertebrates, 54
pigs, 83, 86, 170; placental products from, 25–26, 48; placental type in, 16, 17, 75, 83, 85, 86, 97, 98
pipefish, 52, 53
pithecoid stage of primate placentation, 97–98
placenta accreta, 215
placental functions, 5–6, 43–44, 73–74; nutrient transfer, 12, 25, 31, 36–37, 40, 44; production of information molecules, 5, 9, 12, 20, 22, 45, 78, 88, 142, 156, 163, 166, 225; regulatory, 8–10, 20–21, 44–46, 74, 163–95; variations in, 3–4, 9
placental growth factor, 48, 106, 194
placental lactogens (PLs), 21, 113, 114, 156–58, 167, 168, 184
placental mammals, 2, 3, 13, 14–16, 68–69; phylogeny of, 12–13, 71–72; placental types in, 3, 14, 72, 73–95
placental products, 25–27, 47–49
placental types, 3, 14–18, 20, 82–90, 224; ancestral, 14, 15, 72, 81, 91–93, 97; comparative advantages of, 117–18; endotheliochorial, 88–89; epitheliochorial, 83–88; Grosser's classification of, 40–41, 83; hemochorial, 89–90; phylogeny and, 88, 92–95, 109, 117; synepitheliochorial, 88. *See also specific types*
placentas: ancestral, 14, 15, 72, 81, 91–93, 97; biomedicine of, 11; characterization of, 82–90; development of, 1–2, 153–54; disposal of, 24, 47; eating of, 11, 47–48, 184; embryological view of, 39–40; as endocrine organ, 9, 12, 20–21, 44–46, 48; eutherian, 14–16; evolutionary biology of, 2–3, 4–6, 13, 96–118, 223–24; expulsion of, 24, 31–32, 33, 74; formation of, 1, 12, 15, 19, 74, 76, 151; gene expression by, 15, 46, 80, 97, 141–42, 151–52, 154–62; historical investigations of, 11–12, 24–49; immunology of, 22; infections of, 215–16; information molecules produced by, 5, 9, 12, 20, 22, 45, 78, 88, 142, 156, 163, 166–84, 225; as link between mother and offspring, 1–3, 133–35; magic of, 24, 25–29; physiology of, 43–44; as regulatory organ, 8–10, 20–21, 44–46, 74, 163–95, 225–26; sex and, 119–37; tissues of, 76–81; word for, 34, 35. *See also* human placenta
placentation: comparative, 13–18, 73–95; human, 17, 96–118; lactation preceding, 69; primate, 97–102
placentomes, 16
placentophagy, 11, 47–48, 184
placoderms, 12–13, 50, 64
platypuses, 14, 62, 67, 69, 122–24, 132
polyploidy, 202
porcupines, 14, 85
postpartum period, 27, 47; care of neonate during, 73, 165; marmoset fertility during, 101; maternal depression during, 47–48, 184; maternal hemorrhage during, 226
precocial offspring, 68, 73, 198, 227
preeclampsia, 6, 7, 23, 105, 196, 197, 219–22, 229; in apes, 219, 220; CRH and, 178; immune maladaptation hypothesis of, 220, 222; inflammatory response and, 22, 189, 191; IUGR and, 171, 218, 221, 222, 226; leptin and, 170, 171; obesity and, 187; parent-offspring conflict and, 210; as placental disease, 196, 215; uterine NK response and, 194

pregnancy loss, early, 6, 7, 23, 196, 200–205; as adaptive mechanism, 203–4, 210; CG and, 173; due to chromosome abnormalities, 202–5; leptin and, 170; in macaques, 202; morning sickness as protective against, 23, 203, 207; rate of, in humans, 200, 222, 229; uterine NK response and, 194, 226
preterm birth, 6, 7, 112, 189, 196, 205, 216–19; CRH and, 21, 178, 217, 219; human susceptibility to, 218–19; placental infections and, 215; progesterone and, 178
Priestley, Joseph, 37, 39
primates, 72; fertility of, 197; GH genes in, 21, 93, 114, 155–56, 168–69; nutrition of neonates, 198; placental types in, 3, 14–17, 40, 91; placentation in, 97–102; strepsirrhine, 78, 86, 87, 109. *See also* anthropoid primates; prosimian primates
Profet, Margie, 211, 216
progesterone, 77, 184; CG and production of, 21, 74–75, 88, 173, 204–5; conversion of, to estrogen, 179; CRH-inhibited production of, 178; placental, 179–80, 204; to reduce preterm birth risk, 178; for uterine receptivity, 74–75, 77, 88, 173, 192–93, 204, 209
prolactin (PRL), 21, 104; gene for, 21, 93, 156–57, 167; in osmoregulation, 156, 158–59; receptor for, 158, 167, 168
prosimian primates, 114, 148, 156, 176–77; galagos, 100, 116; placental type in, 3, 17, 91, 93–94
proteomics, 24, 25

rabbits, 23, 42–43, 72, 79, 85, 89
raccoons, 16, 84
radioisotope studies, 11, 43
Ramsey, Elizabeth, 42
rats, 25, 145; CRH in, 176, 177; fatty acid transport in, 108; leptin in, 171, 172; placental gene expression in, 155; placenta type in, 38–39, 85, 89; placentophagy in, 48; prolactin gene in, 156–57; sex chromosomes of, 125, 137; syncytin proteins in, 79
red pandas, 16, 84, 94
regulatory physiology, 2, 4–5, 8–9, 44–46, 163–66, 225
Renaissance, 32–39
Renfree, Marilyn, 62–63
reproductive efficiency, 197–200
reproductive fitness, 3, 17, 19, 23, 134, 164, 197, 198; maternal immune response and, 212, 226; parent-offspring conflict and, 135, 136, 213, 227; paternal, 52, 134, 137, 213; placenta and, 92, 119, 213, 214, 215; pregnancy loss and, 200–201, 209–10
reptiles, 13, 59, 63, 123, 124, 133; eggs of, 2, 12, 53, 55; live birth of, 12, 13, 51, 59, 122; placenta of, 15, 59; sex determination in, 120, 122, 123; squamate, 55, 59, 64, 133; VT genes in, 70
retrotransposons, 142, 144, 146–47, 149
retrovirus(es), 147–49; koala, 150; in mammalian genome, 10, 17, 19–20, 147; in placental tissues, 81, 93, 117, 142, 148–49, 151–52, 194; syncytins from, 79, 148
Rhesus macaques, 116, 117, 157, 176, 177; placental type in, 42, 43, 85, 101–2
rheumatoid arthritis, 195
RNA, 25, 127, 152; genome size and, 143; HLA, 194; messenger (mRNA), 141, 144, 171; micro (miRNA), 141, 144, 145; noncoding genes, 143, 144; placental leptin, 171; retrotransposons and, 146; in syncytiotrophoblast nuclei, 79; transcribed, 79, 127, 141, 144, 150–51; of *Xist* gene, 125, 129, 144
rock hyraxes, 16, 84, 89, 90
rodents, 17, 23, 39, 72, 94, 148; GH gene in, 158, 167; placental gene expression in, 154–55; placental lactogens in, 21, 156–58; placenta of, 14, 16, 42–43, 82, 89, 90, 94, 98; prolactin gene in, 93, 157; sex chromosomes of, 8, 18, 125, 130; syncytin proteins in, 79, 148; syncytiotrophoblasts in, 78–79, 81, 90; TGCs in, 76–77, 78
ruminants, 77, 78, 94, 156; placenta in, 16, 36, 78, 82, 88, 114

salamanders, 53, 58, 124
sauropods, 13, 124
sauropsids, 63–66, 67
sauvagine, 175
Scheele, Carl, 37
scorpions, 50
seahorses, 52–53, 223
seals, 16, 72, 165, 198
seasonality and reproduction, 203
Selenka, Emil, 40

serotonin (5-HT), 183–84
sex chromosomes, 7–8, 18, 120, 122–31; consequences of, 125–26; evolution of, 123–24; heteromorphic, 121, 125–26; human X chromosome, 124–25; of marsupials vs. eutherians, 128–29; X chromosome inactivation, 126–31
sex determination, 7, 18–19, 119–22; live birth and, 121–22; sex chromosomes and, 122–24; vs. sex differentiation, 120; temperature-dependent, 120–21, 122
sharks, 44, 50, 60–61
sheep, 156, 168; CRH in, 173, 176, 177; leptin in, 170, 172; placental type in, 16, 34, 36, 84, 88, 91
siamangs, 181, 228
skinks, 59, 120, 121
sloths, 16, 72, 88, 94
small-for-gestational-age fetuses, 23, 171, 186, 201, 215
Smith, Homer, 158
snakes, 13, 59, 61, 66; sex chromosomes of, 18, 122, 123, 127
somatostatin, 167
sperm, 1, 2, 132, 134, 212, 222; of angiosperms, 133; fertilization of eggs by, 52, 154; pathogens borne on, 122, 216; sex chromosomes of, 18, 121, 122, 126, 129; of sharks, 60; swallowing of, 220
spiral arteries, maternal, 37; CG and, 173; extravillous cytotrophoblasts and, 172, 173, 213–14, 215, 228; human, 102–3; macaque, 101–2; remodeling of, 103–5, 110, 112, 117, 161, 172, 173, 218, 220, 221, 227, 228; TGCs associated with, 76, 77; tropoblast plugging of, 102, 105, 106, 112
stem cells, 27, 45, 80, 151
steroids, 5, 9, 21, 25, 45; adipose tissue production of, 186–87; adrenal, 174, 179; in bird eggs, 2; estrogens, 21, 45, 114, 179, 180, 184
stillbirth, 29, 187, 202, 205, 215
Strahl, Hans, 40
synapsids, 13, 63–66, 124, 223, 226
syncytia, 17, 78, 79, 81, 83, 88, 100
syncytin proteins, 79, 148
syncytiotrophoblast membrane microparticles (STBMs), 189
syncytiotrophoblasts, 17, 40, 78–81, 89, 90, 100, 172; ages of nuclei in, 80–81; exosomes released by, 194–95; formation of, 78–79; gene expression in nuclei of, 80; GH2 secretion by, 167; human, 78, 104, 114; immune barrier function of, 216; macaque, 101, 114; marmoset, 101; in maternal IgG transfer to fetus, 115, 117; peptides secreted into maternal circulation by, 79–80; in regulation of placental signaling, 79–80
syndesmochorial placentas, 41
synepitheliochorial placentas, 14, 41, 77, 88

tapirs, 72
tarsiers, 93, 98, 160, 177, 181; placental type in, 3, 17, 85, 100
tarsioid stage of primate placentation, 97
technological advances, 24–25, 40, 43, 44, 46, 138, 225; IVF, 153–54
tenrecs, 16, 84, 89, 99, 111
"third brain," placenta as, 20
"thrifty phenotype hypothesis," 185–86
thyroid stimulating hormone (TSH), 159–60, 172, 173, 208
thyrotoxicosis, transient, 173, 208
toads, 58
Toxoplasma gondii, 215
transposons, 138, 140–41, 142, 144, 145–47
tree shrews, 72, 88–89, 94
trisomies, 202; trisomy 21, 204–5
Trivers, Robert, 135, 136
trophoblast giant cells (TGCs), 76–77
trophoblasts, 15, 76; bipedalism and, 109; CG and, 161–62, 204; CRH regulation of, 176, 178; of endotheliochorial placentas, 89; of epitheliochorial placentas, 83; formation of, 130, 154; of hemochorial placentas, 89–90, 112; horse, 87, 99; human, 102–5, 106, 109; interaction of, with maternal immune system, 166, 193–94, 220; invasion of decidua by, 82–83, 91, 109, 112, 117, 176, 213–15, 221, 225–26, 228; mammalian placental morphology and, 84; marsupial, 130, 154; in maternal-placental signaling, 113, 114; paternal X chromosome inactivation in, 130; plugging of maternal spiral arteries by, 102, 105, 106, 112; postpartum maternal hemorrhage and, 226; preeclampsia and, 220–21; primate, 43, 101, 228; rodent, 130; term for, 40

tuataras, 64, 66, 123
turtles, 13, 55, 64, 66, 120, 123

umbilical cord, 15, 28, 29, 32, 35, 51, 54, 61, 154; Wharton's jelly within, 44–45
urocortins, 173, 175
uterine milk, 9, 61
uterine natural killer (uNK) cells, 193–95, 212, 216, 225; preterm birth and, 220–21
uterine paps, 31, 35
uterus, 12, 68, 134; bipedalism and, 109; blastocyst invasion of wall of, 3, 77, 82–83, 213–15; CRH expression by, 176; decidua of, 77, 103, 192, 212, 214; early illustrations of, 34, 37; and exposure to pathogens, 211–12, 229; expulsion of placenta from, 24, 31–32, 33, 74; immune response and, 22, 46, 192–94; menstruation and, 211, 213, 222; monotreme, 82; and preparation for pregnancy, 192–93; progesterone and receptivity of, 74–75, 77, 88, 173, 192–93, 204, 209; rodent, 77; shark, 60

vascular endothelial growth factor (VEGF), 106, 188, 194
vertebrates, 3, 78; CRH in, 173–75; evolutionary history of mammals, 63–69; gene imprinting in, 3, 19, 132–33; live birth of, 50, 51; phylogeny of, 54, 63; placental function and evolution in, 10, 21, 42; placental gene expression in, 154–55; regulatory physiology of, 164; sex determination in, 7, 120–21; tetrapod, 13, 53; transposons in, 145–47
Vesalius, Andreas, 34, 35
vitellogenin (VT) genes, 70
viviparity, 46, 50, 58, 59, 61, 137. *See also* live birth
von Haeckel, Ernst, 91

Waddington, C. H., 131
Waterland, Robert, 145
W chromosome, 18, 122–23, 125
weaning, 10, 198, 199, 200
weight gain, gestational, 184, 190
whales, 16, 72, 98, 227
Wharton, Thomas, 44
Wharton's jelly, 44–45

X chromosome, 8, 18, 122–25; human, 124–25; mosaic expression of, 18, 126–27, 128, 130
X chromosome inactivation, 8, 18–19, 126–31; genes that escape, 130–31; in marsupials vs. eutherians, 128–30; paternal, 8, 18, 128, 129–30, 132, 137
Xenarthra, 71, 72, 84, 86, 94
Xist gene, 125, 129–30, 144

Y chromosome, 18, 122–25, 129
Yen, Samuel, 20
yolk sac, 12, 15, 19, 31, 50, 51, 60, 76, 82, 151; marsupial, 62; persistent, 76, 97; secondary, 102, 104; shark, 60, 61; tadpole, 58
yolk-sac placentas, 61, 62, 69, 71, 76, 77, 82, 86

Z chromosome, 18, 122, 123–25, 127
zona pellucida, 154
zonary placental type, 15–16, 34; endotheliochorial, 14, 16, 84, 89; hemochorial, 84, 94
zygote, 154